NISTIR 7841

Study of Early-Age Bridge Deck Cracking in Nevada and Wyoming

Dale P. Bentz
Paul E. Stutzman
Aaron R. Sakulich
W. Jason Weiss
Materials and Structural Systems Division
Engineering Laboratory
National Institute of Standards and Technology
Gaithersburg, MD 20899

Prepared for Federal Highway Administration Central Federal Lands Highway Division

I0494048

January 2012

U.S. Department of Commerce
John Bryson, Secretary

National Institute of Standards and Technology
Patrick D. Gallagher, Director

Abstract

In late 2009, the Echo Wash and Valley of Fire bridge decks were constructed in the Lake Mead National Recreation area in Nevada. Within six months after installation, in early 2010, both decks exhibited considerable transverse cracking, with some cracks extending through the thickness of the deck. Similar cracking was observed in the Snake River bridge deck in Wyoming. This report details the results of a two-pronged approach to examining the causes of such cracking. First, for the Nevada bridge decks, materials similar to those used in the construction of these bridge decks were obtained and mortars were prepared and evaluated for chemical shrinkage, autogenous shrinkage, and drying shrinkage. Second, cores from all three bridge decks were obtained and analyzed using both optical microscopy and scanning electron microscopy to assess air contents, paste and aggregate volume fractions, and the overall nature of the concrete microstructure. In all three cases, the concrete mixtures had been reportedly batched at a considerably lower water-to-cementitious materials ratio (w/cm) than that in the approved concrete mixture proportions, with reductions from the specified w/cm of approximately 0.40 to w/cm of 0.35 (Snake River) to 0.31 or 0.32 (Nevada). In the laboratory mortars, w/cm was therefore varied between 0.30 and 0.40 to examine the influence of this variable on shrinkage and cracking. The observed effects of w/cm on shrinkage and cracking were small, while petrographic analysis revealed that the in place concrete cores exhibited a w/cm of 0.4 or slightly higher, consistent with the concretes having been retempered with additional water after being batched. Such retempering could produce an inhomogeneous microstructure with dramatically high air contents that could ultimately contribute to the observed cracking. The high air contents were not detected in either the fresh air content determinations performed on site or the prepared compressive strength cylinders, likely due to the removal of entrapped air during test sample preparation in both cases. For the Nevada bridge decks, the cracking occurred following a dry period where average daily ambient relative humidities fell below 30 %, greatly increasing the drying stresses and likely providing a final contribution to the observed cracking.

Keywords: Air content; autogenous shrinkage; bridge deck; early-age cracking; petrography; retempering.

Table of Contents

List of Figures

List of Tables

1. Introduction

It is often considered inevitable that infrastructure materials such as concrete will undergo cracking, a complex phenomenon that can significantly affect the service life of a structure [1]. Cracks may degrade bulk mechanical properties, may allow aggressive media such as chlorides easy access to reinforcing steel, and may propagate over time increasing the damage to a structure [2].

Late-age cracking can occur for a variety of reasons. Mechanical cracking can occur due to structural settlement, excessive loading, and freeze/thaw cycling. Cracking that occurs due to alkali-aggregate reactions (whether alkali-silica reactions or alkali-carbonate reactions), the corrosion of reinforcing steel, or sulfate attack (whether internal or external) have a more chemical origin, as expansive reaction products are formed that create internal forces, leading to cracking [2]. These reactions, however, generally take place over the course of years in mature concrete systems and are not considered to be viable causes for the cracking observed in the bridge decks examined in the current study.

Contributors to early-age cracking of concrete include plastic shrinkage and settlement, drying shrinkage, autogenous shrinkage, and thermal stresses [2, 3]. Drying shrinkage occurs when water evaporates from the concrete to the environment, creating stresses in the concrete via the remaining (partially) filled water menisci. Autogenous shrinkage, sometimes considered to be a special case of drying shrinkage [2], occurs when water-filled menisci are created not by transfer of water to the external environment but by the inherent chemical shrinkage that accompanies the cement hydration and pozzolanic reactions [3]. Chemical shrinkage occurs due to the fact that the final hydration products in cement have a smaller volume than the initial reactants (including water) [4, 5]. Autogenous and drying shrinkage at early ages can be compounded by thermal stresses that are created as the concrete heats up and subsequently cools, due to both the daily temperature cycle and the substantial heat generated by the exothermic cement hydration and pozzolanic reactions. It is often the superposition of these thermal and shrinkage stresses that ultimately causes the cracking of a concrete at a young age [3].

Regardless of mechanism, cracking is the largest contributor to the degradation of infrastructure and thus has a substantial impact on the national economy [6]. Approximately $20 billion is spent yearly on rehabilitation, maintenance, and protection (waterproofing, etc.) of American infrastructure [7], yet the American Society of Civil Engineers gives American Infrastructure a grade of 'D' and estimates that $2.1 trillion should be spent before 2014 [8]. Nationwide, some 4,000 dams are considered deficient (of which 1,819 are considered 'high hazard'); about 27 billion liters (7 billion gallons) of clean drinking water are lost daily to leaking through cracked pipes; 11 % of the nation's bridges are structurally deficient; and one-third of major American roads are in poor condition [8, 9].

Despite extensive efforts by numerous state Departments of Transportation to produce concrete that doesn't crack at early ages, detrimental early-age cracking continues to occur, as witnessed by several major cases in the past few years. For example, the Valley of Fire and Echo Wash bridges (Northshore Road) in Lake Mead National Recreation Area exhibited early-age cracking within six months of their completions on October 30 and November 6, 2009, respectively. Within six months of their casting, 133 cracks were documented on the topside of the Valley of Fire bridge deck, with 91 cracks on the underside, while 150 cracks were documented on the topside of the Echo Wash bridge deck, with 105 cracks on the underside [10]. These cracks were linear, oriented transverse to the bridge deck, ranged from 0.025 mm to 1.14 mm in width, and ranged from 0.25 m to 6.9 m in length. One of the cracks on the underside of the Echo Wash Bridge was observed to run completely across the bridge deck. In 2009, another bridge in Grand Teton National Park on the Snake River Bridge Project, WY PRA GRTE 13(8), also experienced early age cracking similar to the bridges on the Northshore Road Project.

Initial examination of the documentation relating to the bridges showed that although the initial submitted mixture proportions called for a water:cementitious materials ratio (*w/cm*) of 0.44 by mass, the approved mixture proportions used a *w/cm* of 0.40, and the concrete was likely batched with *w/cm* of about 0.32 and 0.31 for the Valley of Fire and Echo Wash bridges, respectively [10]. In the case of the bridge deck in Wyoming, the *w/cm* was estimated to be approximately 0.35, versus 0.40 for the approved mixture proportions. Additional water could have presumably been supplied close to placement time at the job sites; however, batch tickets contained no information about the occurrence of any such water additions. The mixture proportions for the Northshore Road bridges also called for a reddish-brown pigment, high- and normal-range water reducing admixtures, an air entraining agent, and a stabilizer/hydration retarder. The weather during the construction time period in Nevada (October to November 2009) was temperate and considered conducive to the pouring and placement of concrete.

In January 2010, a private engineering firm was called in to perform inspections of the two Northshore Road bridges using Impact Echo (IE) and Spectral Analysis of Surface Waves (SASW) analyses, two non-destructive characterization methods that measure waves propagating through the bridge deck itself to identify defects. The firm concluded that the Valley of Fire bridge had no areas of significant defects and was structurally sound; the Echo Wash bridge was also structurally sound, although there were some areas of "surface distress", unlikely to extend deeper than 50 mm, that were primarily an aesthetic, and not structural, concern [11, 12].

In the months immediately following these late January 2010 assessments, both bridge decks began to exhibit substantial transverse cracking. Assured by the non-destructive testing that the bridges themselves were structurally sound, FHWA contracted with a private laboratory to perform a petrographic examination of cores taken from the cracked bridge decks, so as to determine the cause of the observed early-age cracking [13]. This examination determined that the composition of the concrete was consistent with that indicated by the mixture proportions and that the microstructure was more representative of cement paste with a *w/cm* of 0.4, rather than a *w/cm* of ≈ 0.3. The petrographer also reported clear evidence of late water addition (dense hydration products around cement grains with a low *w/cm*, while the bulk paste was more porous, mottling of paste color, and a high air content indicative of retempering). Furthermore, substantial air voids were present, giving the concrete a 'Swiss cheese' appearance and an estimated air content of ≈ 15 %, compared to the ≈ 6 % called for in the mixture proportions and the ≈ 8 % reported by field testing on the fresh concrete. The author of that report came to the conclusion that there was no evidence of autogenous shrinkage (in the form of randomly oriented microcracks evenly distributed throughout the interior of the samples), while the observed oriented surface cracking is commonly associated with drying shrinkage. Further, the presence of numerous bleed voids indicated that plenty of water was available to the hydrating system, supporting the belief that self-desiccation (and thus autogenous shrinkage) was unlikely to have occurred.

Although conclusions have been drawn from this petrographic examination, no study has yet observed the actual shrinkage and cracking process in the materials used in these two bridge decks. This study therefore uses materials similar to those used in the Valley of Fire and Echo Wash bridges to create mortar mixtures with differing *w/cm* of 0.30, 0.35, and 0.40. Chemical shrinkage, autogenous shrinkage, and drying shrinkage tests were carried out to quantify their potential impact on cracking, while a limited number of restrained ring shrinkage specimens were fabricated and observed for cracks. Extensive petrographic examinations were also carried out, with additional analysis being provided by scanning electron microscopy analysis, a technique that was not included in the original petrography report [13].

2. Preliminary Observations

2.1 Cylinder Strengths vs. Air Contents

FHWA provided summaries of the 28-d compressive strength testing data of cast cylinders for the concretes for each bridge deck, along with their fresh air contents [10]. For both bridge decks, the specified 28 d strength was 31 MPa. These results were plotted in each case and are provided in Figure 1 and Figure 2 for the Valley of Fire and Echo Wash concretes, respectively. There is a significant amount of scatter in both data sets, but the best fit regression lines in both cases indicate about a 5 MPa decrease with each 1 % increase in air content, consistent with the U.S. Bureau of Reclamation general rule of thumb that a 1 % increase in air content will decrease compressive strength by about 5 % [14]. **The lone strength point taken later from drilled cores (June 2010) from the Valley of Fire bridge deck is consistent with the overall trend of strength vs. air content data, considering the in-place air content to be on the order of 13 %.** These initial plots raised two important points for further study. First, it is important to note that while an increase in air content decreases compressive strength, a decrease in *w/cm* significantly increases it. For the Valley of Fire and Echo Wash bridge deck concretes, thus, the normal consequence of not meeting specified strengths in the case of abnormally high air contents could have been effectively masked by the concurrent reduction in *w/cm*, if present. Second, there is also the question of whether the preparation of the cylinder specimens in the field (specifically their rodding and compaction) might effectively remove some of the air (most likely the entrapped air) from the concrete, producing cylinder test specimens with significantly lower air contents and higher strengths than those of the in-place concrete. This will be discussed further later in this report.

2.2 Weather Observations

Data from the weather station in Overton, NV for the time period from the casting of the bridge decks (October 2009) until the observation of significant cracking (February to March 2010) are summarized in Figure 3 [15]. It is assumed that the cracking mostly occurred sometime in the months of February and March, as it was not noted in the Olson Engineering reports, whose surveys were conducted on January 28-29, 2010. This would suggest that the cracking likely occurred following a "rainy" season that was characterized by high daily average relative humidities in the range of 70 % to 80 %. Following this, the RH steadily decreases to 20 %. This drying trend may have introduced additional stresses that could have initiated the cracking observed in these concretes.

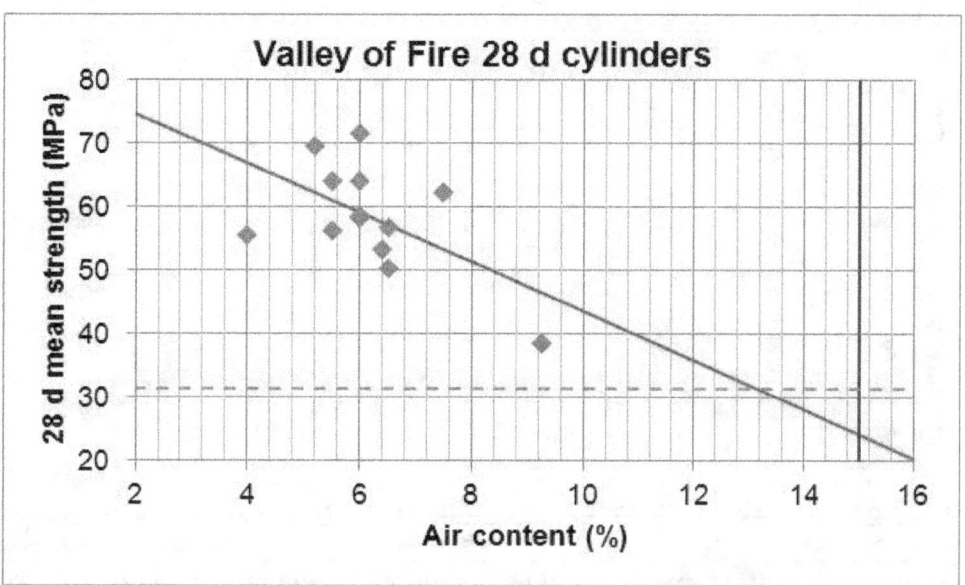

Figure 1. Measured 28 d mean compressive strength vs. air content for Valley of Fire bridge deck concrete [10]. The solid red line indicates the best fit linear equation. The dashed green horizontal line indicates the strength level measured for drilled cores taken in June 2010 [10]. The solid blue vertical line indicates the air content measured in the original petrography report [13].

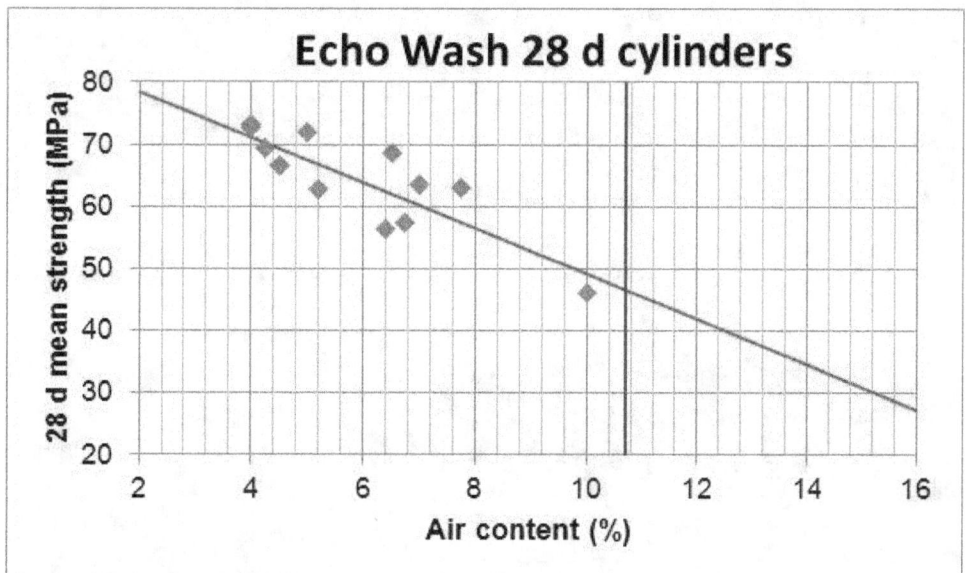

Figure 2. Measured 28 d mean compressive strength vs. air content for Echo Wash bridge deck concrete [10]. The solid red line indicates the best fit linear equation. The solid blue vertical line indicates the air content measured in the original petrography report [13].

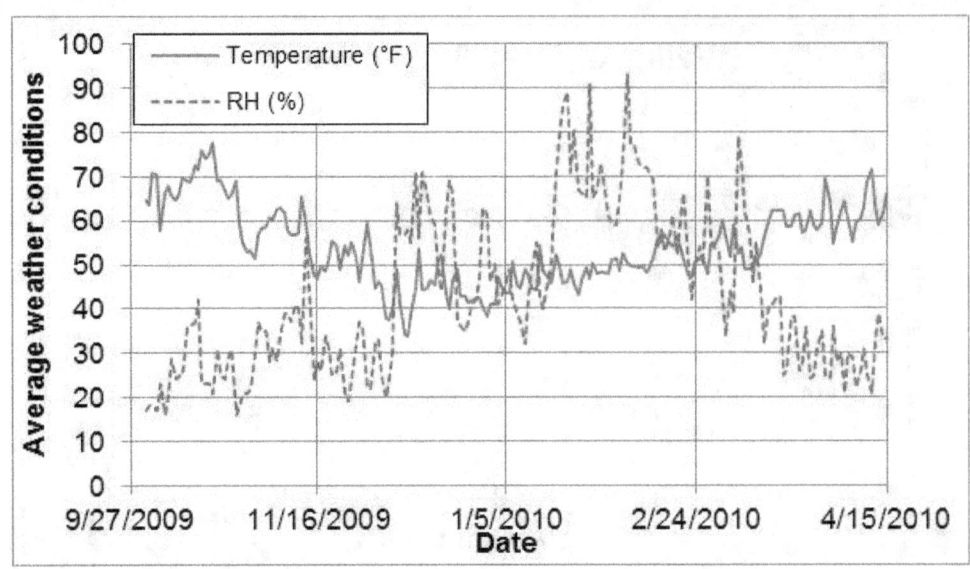

Figure 3. Average daily air temperature and ambient relative humidity in Overton, NV, for the time period from 10/1/2009 until 4/15/2010 [15].

3. Task 1.

The following is an excerpt from the agreed upon statement of work between NIST and FHWA, specifying the items that were to be addressed in Task 1.

A. **Task 1** – Conduct autogenous strain testing in accordance with ASTM C 1698-09, "Standard Test Method for Autogenous Strain of Cement Paste and Mortar" using mixture materials and proportions representative of the Northshore project bridges. Perform three sets of testing (3 mortar specimens each) at water-cementitious materials ratios by mass of 0.30, 0.35, and 0.40. Conduct drying shrinkage testing of these same mortars following the guidelines in ASTM C 596-09, "Standard Test Method for Drying Shrinkage of Mortar Containing Hydraulic Cement." Mortar mixtures or their paste components will also be evaluated with respect to chemical shrinkage by ASTM C 1608-07, "Standard Test Method for Chemical Shrinkage of Hydraulic Cement Paste," isothermal calorimetry (following the guidelines of ASTM C1702-09a, "Standard Test Method for Measurement of Heat of Hydration of Hydraulic Cementitious Materials Using Isothermal Conduction Calorimetry) and semi-adiabatic calorimetry, the latter to assess the potential for a significant contribution of thermal stresses and strains to the observed cracking.

3.1 Materials and Experimental Procedures

Other than water for the mixing of pastes and mortars, all materials employed in this project were received via the Central Federal Lands Highway Division. These included the cement, fly ash, pigment, fine and coarse aggregates, and four chemical admixtures (Sika AE air entraining agent, Delvo stabilizer, Plastiment water reducing admixture and Sika Viscocrete high range water reducing admixture[1]). The cement is an ASTM C150 Type II/V, low alkali cement with a Blaine fineness of 363 m^2/kg; its oxide composition is provided in Table 1. Its estimated Bogue composition is 61 % tricalcium silicate, 15 % dicalcium silicate, 4 % tricalcium aluminate, and 12 % tetracalcium aluminoferrite by mass. A density of 3160 kg/m^3 was measured for the cement powder using ASTM C188 [16], with values of 3150 kg/m^3 and 3170 kg/m^3 being obtained for two replicate specimens. This is in general agreement with the value of 3150 kg/m^3 utilized by the concrete producer. The oxide composition of the Class F fly ash as reported by its supplier is provided in Table 1. The fly ash provides ASTM C311 [16] strength activity indices of 87.8 % and 95.3 % at 7 d and 28 d, respectively. For the fly ash, two density measurements (ASTM C188) yielded values of 2220 kg/m^3 and 2230 kg/m^3, slightly lower than the value of 2310 kg/m^3 included by the supplier on its specification sheet. The pigment is a red powder, based on iron oxide.

The particle size distributions (PSDs) of the cement, fly ash, and red pigment were measured using a laser-diffraction instrument with isopropanol as the dispersant. The measured PSDs are provided in Figure 4. The pigment is finer than the other powders, with a median diameter of less than 1 μm, and it was hypothesized that such a fine additive may negatively

---

[1] Certain commercial products are identified in this paper to specify the materials used and procedures employed. In no case does such identification imply endorsement or recommendation by the National Institute of Standards and Technology, nor does it indicate that the products are necessarily the best available for the purpose.

impact both the workability/water demand and the shrinkage, particularly autogenous shrinkage, due to its high surface area and reduced interparticle spacing. For this reason, a fourth mortar with a water-to-cementitious materials ratio (*w/cm*) of 0.30 by mass without any pigment was added to the originally requested series of three mortars with pigment and *w/cm* of 0.30, 0.35, and 0.40. In Figure 4, the composite curve indicates the calculated overall powder PSD, assuming a blend with 67 % cement, 31 % fly ash, and 2 % pigment by volume (the mixture proportions employed in prepared mortars). While most of the testing was conducted on mortar specimens, preliminary measurements of chemical shrinkage were performed on pastes to assess the contributions of the fly ash to the early age reactions. Cement pastes with and without fly ash (*w/cm* = 0.40) were prepared and their chemical shrinkage evaluated according to the ASTM C1608 standard test method [16]. These pastes, along with an additional *w/cm* = 0.40 paste with pigment, were also evaluated for their initial and final times of set using the ASTM C191 standard test method (Vicat needle penetration) [16].

Table 1. Oxide compositions (mass percentages) of cement and fly ash

Oxide	Cement	Fly Ash
CaO	64.4 %	6.4 %
SiO_2	21.1 %	58.3 %
Al_2O_3	3.9 %	22.5 %
Fe_2O_3	3.8 %	5.1 %
MgO	1.0 %	not reported
SO_3	3.3 %	not reported
Equivalent alkali as Na_2O	0.54 %	not reported

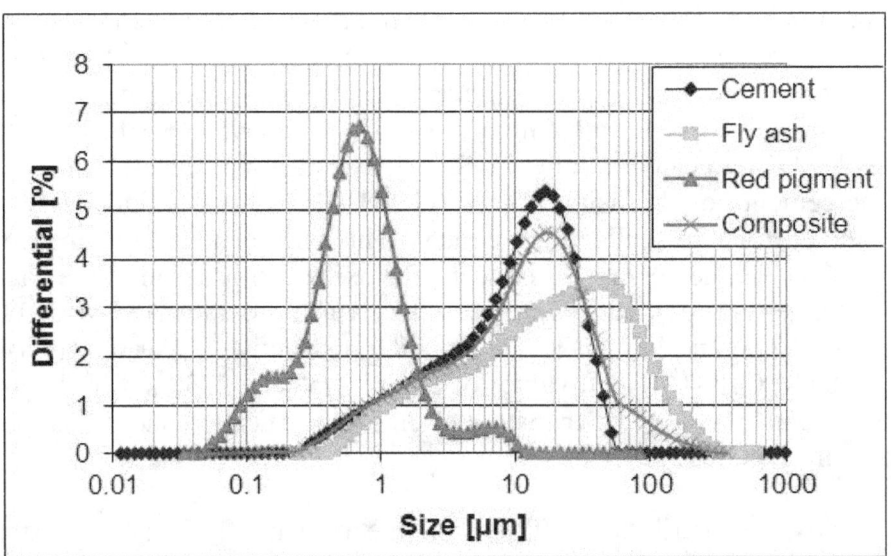

Figure 4. Measured particle size distributions for cement, fly ash, and pigment and that computed for the blend investigated in the present study.

The mortar mixture proportions were designed by eliminating the coarse aggregate from the approved concrete mixture proportions. It was assumed that admixture absorption by the coarse aggregates would be minimal (due to their much smaller surface area relative to the fine aggregates and the blended cementitious powder) and therefore the admixture dosages per unit mass of cement were held constant in going from concrete to laboratory mortars. The base mixture was prepared with a w/cm = 0.40. Since the delivered concrete batch tickets indicated a reduction in this w/cm, to somewhere in the range of 0.32, mortars were also prepared with w/cm of 0.35 and 0.30, by reducing the water in the mixture by the appropriate amount, while maintaining the ratio of aggregate to cementitious powders constant. The admixture dosages were not adjusted for the w/cm = 0.35 mortar, but for the two w/cm = 0.30 mortars (with and without pigment), an increase of the Sika Viscocrete was deemed necessary to produce a workable mixture for molding cubes for strength, prisms for drying shrinkage, and corrugated tubes for autogenous shrinkage measurements. The complete mortar mixture proportions on a mass per batch basis are provided in Table 2. The w/cm = 0.40 mortar was prepared first, and subsequent batch sizes were increased to assure an adequate volume of mortar for preparing the various required specimens. Table 2 indicates the exact additional dosages of the Sika Viscocrete that were necessary to produce a mortar with adequate workability for casting cubes, prisms, and corrugated tubes for the two w/cm = 0.30 mortars.

Table 2. Mortar mixture proportions and fresh properties

Ingredients/Designation	FHWA 0.4	FHWA 0.35	FHWA 0.3	FHWA 0.3np
Cement	1668 g	1918 g	1970 g	1989 g
Fly ash	558 g	641 g	659 g	665 g
Pigment	67 g	77 g	79 g	0
Water	887 g	896 g	789 g	796 g
Sand	4314 g	4960 g	5095 g	5142 g
Sika AE agent	0.37 g	0.43 g	0.44 g	0.44 g
Sika Viscocrete	3.92 g	4.50 g	4.63 g + 5.28 g	4.67 g + 5.44 g
Plastiment	1.28 g	1.47 g	1.51 g	1.52 g
Delvo stabilizer	1.17 g	1.34 g	1.38 g	1.39 g
w/cm[1]	0.4	0.35	0.3	0.3
Mix Temperature	24 °C	24 °C	25 °C	25 °C
Air content[2]	7.6 %	5.1 %	6.3 %	5.9 %

[1] Cementitious material is composed of cement and fly ash only (no pigment) for purposes of computing w/cm.
[2] Air content computed based on unit mass (cup) measurements and reported specific gravities of materials.

The fresh mortars were characterized with respect to unit weight (air content) and temperature as reported in Table 2. Six prisms (25 mm x 25 mm x 305 mm) were prepared for drying shrinkage (ASTM C596 [16]), nine cubes (50 mm) were prepared for measurement of compressive strength (via ASTM C109 [16]), and 3 corrugated tubes (≈415 mm in length) were prepared for measurement of autogenous shrinkage (ASTM C1698 [16]). In addition, one small sealed vial of mortar (≈8 g of mortar) was prepared for isothermal calorimetry measurements (ASTM C1702 [16]) conducted out to 14 d and one cylindrical specimen (≈380 g of mortar) was cast for semi-adiabatic calorimetry (temperature) measurements to 3 d. Finally, when extra mortar was still remaining, two restrained ring shrinkage specimens (25 mm thick, inner diameter of 100 mm, outer diameter of 150 mm) were cast. For these specimens, the inner restraining ring consisted of a 25 mm thick, 100 mm diameter cylinder of stainless steel.

The semi-adiabatic temperature was measured using a custom-built semi-adiabatic calorimeter unit [17]; replicate specimens from separate batches have indicated a standard deviation of 1.4 °C in the maximum specimen temperature achieved during a 3 d test. Mortar cube compressive strengths were measured at 1 d, 7 d, and 28 d on specimens demolded after 1 d and cured in a saturated calcium hydroxide solution, according to the procedures in ASTM C109 [16], but with a loading rate of 20.7 MPa/min, switching to deformation control (at the instantaneous deformation rate) once a stress of 13.8 MPa was reached; three specimens prepared from a single batch were evaluated at each time, with the averages and coefficients of variation provided in the results to follow.

Autogenous shrinkage was measured on duplicate or triplicate sealed mortar specimens. In the ASTM C1698 standard [16], the single laboratory precision is listed as 28 μm/m for mortar specimens; in general, the measured standard deviations were well in line with this reported precision. Drying shrinkage specimens were demolded after 1 d, placed in saturated calcium hydroxide solution until an age of 3 d, and then exposed to the required 23 °C, 50 % RH environment. Both their length and mass were measured, initially at 3 d and subsequently periodically out to an age of 56 d. At 56 d, the specimens were transferred to a 23 °C, 25 % RH environment and their further shrinkage and mass loss measured after 14 d. Finally, the specimens were placed in a 100 °C oven and dried to constant mass. These final masses, along with their accompanying final lengths (shrinkage), were measured. At ages out to 28 d, in addition to measuring their mass and length, the resistivity of the drying shrinkage mortar prisms was also determined using a 4-point measurement, to examine whether a relationship might exist between increasing drying shrinkage and changes in resistivity, as the pore structure of the specimens progressively dries out.

3.2 Results and Discussion

3.2.1 Chemical Shrinkage

Chemical shrinkage measurements were conducted on paste specimens prepared with only the cement (w/c = 0.40) and prepared with the cement/fly ash proportions (w/cm = 0.40) used in the bridge deck concretes. The plot in Figure 5 indicates that in the cement/fly ash blend, the cement has the dominant effect on the chemical shrinkage until an age of about 7 d, as the measured chemical shrinkage is basically equivalent to that measured for a pure cement paste scaled to its 75 % mass content in the cement/fly ash blend. Beyond that age, a substantial contribution to the reactions is provided by the (pozzolanic) reaction of the fly ash with the cement hydration products, as the curves for the cement/fly ash blends begin to greatly exceed that of the pure cement paste scaled at 75 % at about 7 d. This reaction continues at a fairly high rate out through 28 d, the end of the chemical shrinkage measurement. This enhanced overall reactivity and increased chemical shrinkage at intermediate ages could contribute to additional autogenous shrinkage within the same time period [18].

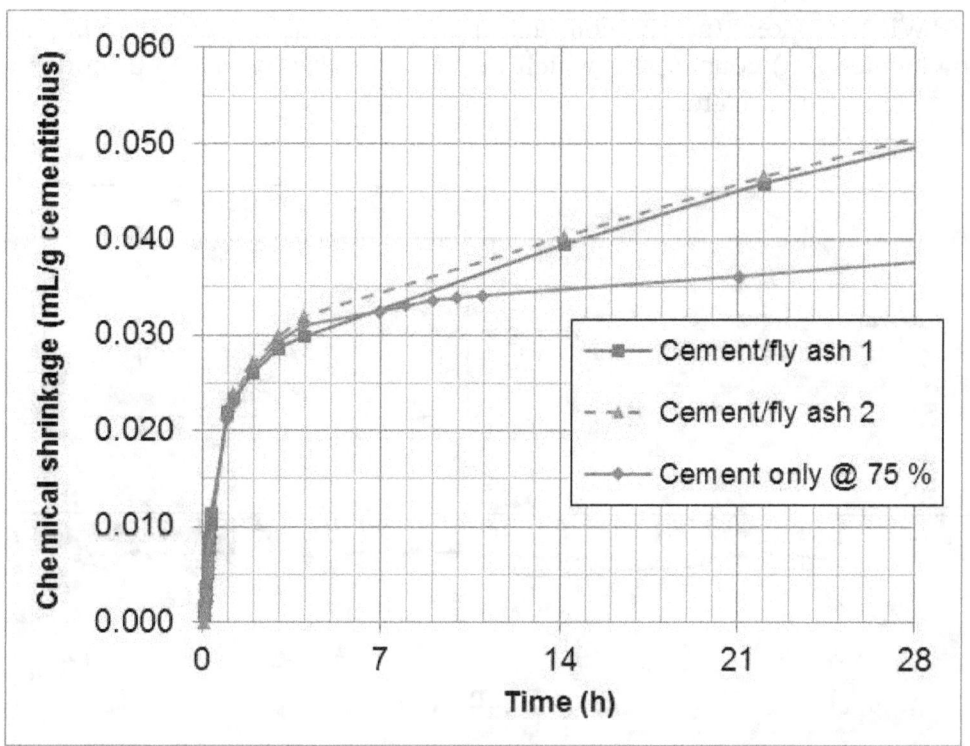

Figure 5. Measured chemical shrinkage for two replicate pastes prepared with cement/fly ash and the scaled (75 %) chemical shrinkage measured for paste prepared with cement only.

3.2.2 Paste Setting Times

Three pastes were examined for their initial and final setting times according to ASTM C191 [16], namely cement only, cement/fly ash, and cement/fly ash/pigment. The results in Table 3 indicate a retardation on the order of 1 h when fly ash or fly ash and pigment are added to the base mixture. This is mainly due to the dilution of the reactive cement component in the latter two mixtures, and likely would not be a source of concern for concretes in the field.

Table 3. Setting times for pastes with w/cm = 0.40

Setting/Mixture	Cement	Cement/fly ash	Cement/fly ash/pigment
Initial Set	2.8 h	3.1 h	3.9 h
Final Set	4.3 h	5.5 h	5.3 h

3.2.3 Calorimetry Results for Mortars

3.2.3.1 Semi-adiabatic Calorimetry

Figure 6 shows the measured temperature for individual specimens from the four mortar mixtures during their first three days of hydration under semi-adiabatic conditions. As would be expected, due to their higher cement content and lower heat capacity (since water has a much higher heat capacity than cement or sand), the lower w/cm systems produce a slightly higher temperature rise. A slight retardation is seen in the w/cm = 0.30 mortar without pigment relative to its pigment-containing counterpart, most likely due to a slight overdosage of the high range water reducer (HRWRA) in the former mixture. For the fairest comparison of performance, the

11

dosage of HRWRA was kept (nearly) constant in going from the *w/cm* = 0.30 pigmented mortar to its non-pigmented (np) counterpart, which in the latter case produced a mortar with greatly increased flow and a small retardation.

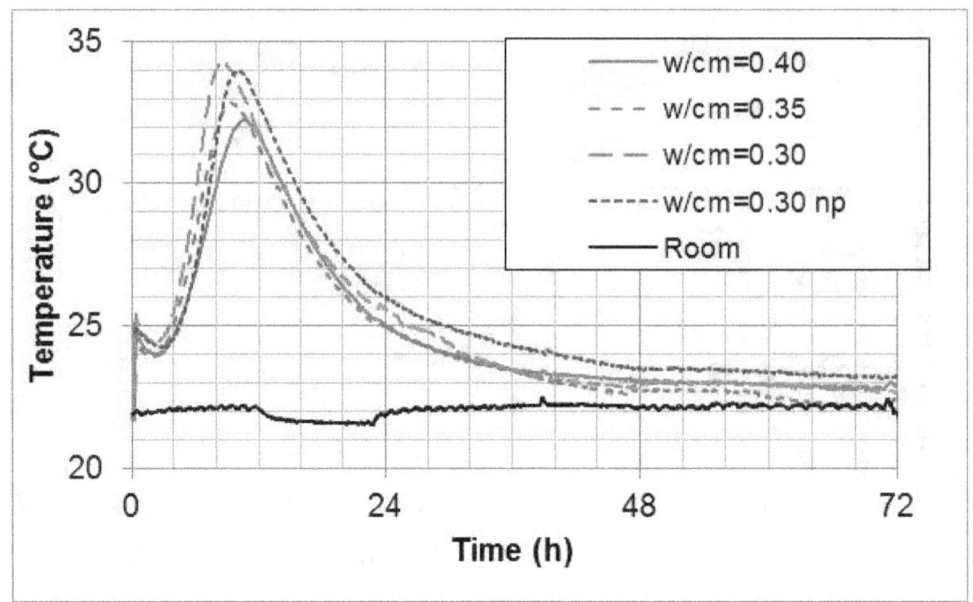

Figure 6. Measured semi-adiabatic temperature rise for the four mortar mixtures.

For plain Portland cement systems, temperature rise is usually highest for an intermediate *w/c* (~ 0.35), as lower *w/c* may not provide sufficient water to maximize hydration and temperature rise [19]. However, for these blended cements with about 25 % fly ash by mass, the *w/cm* of 0.30, 0.35, and 0.40, correspond to equivalent *w/c* of about 0.40, 0.47, and 0.53, respectively. Since the fly ash is fairly inert at early ages, none of these mixtures should be water limited and therefore, the semi-adiabatic temperature response is dominated by the increasing cement content of the lower *w/cm* mixtures. Based on the 25 % fly ash content and the prevailing environmental conditions (temperature) recorded during the casting of the concrete bridge decks, the impact of a potential slightly increased temperature rise, due to the potentially lower *w/cm* of the field concrete mixture, on any likelihood for thermal cracking is deemed minimal. Additionally, such thermal cracking would normally manifest itself much sooner than the approximately four month delay from casting to the observation of considerable cracking for the bridge decks in question.

3.2.3.2 Isothermal Calorimetry

The isothermal calorimetry results for the four mortars are provided in Figures 7 to 9. Figure 7 shows the heat flow results for the first 24 h of hydration. The four curves are quite similar, with a slight retardation observed for the *w/cm* = 0.30 mortar without pigment, once again due to a slight overdosage of the HRWRA. Figure 8 shows the cumulative heat release curves, with the expected effect that higher *w/cm* mortars produce greater heat release at later ages, due to the self-desiccation that occurs and somewhat limits further hydration in the lower *w/cm* mortars. The curves for the four mortars begin to diverge after approximately 18 h of hydration. The cumulative heat release curves for the *w/cm* = 0.30 mortars, with or without pigment, are basically identical, indicating that the initial hydration retardation occurring in the mortar without pigment had minimal if any lasting impact on the hydration and pozzolanic reactions.

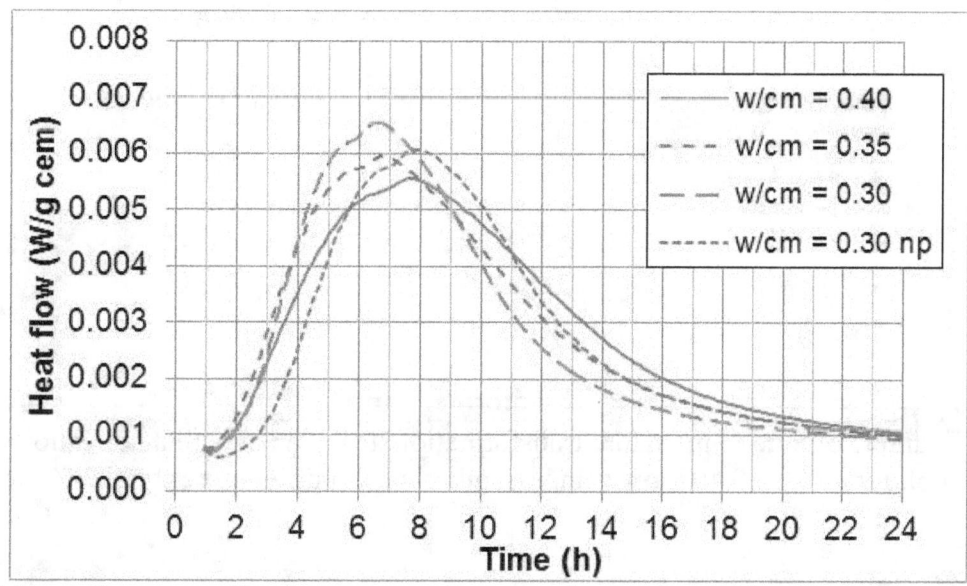

Figure 7. Heat flow normalized per mass of cement for the first 24 h of hydration for the four mortar mixtures.

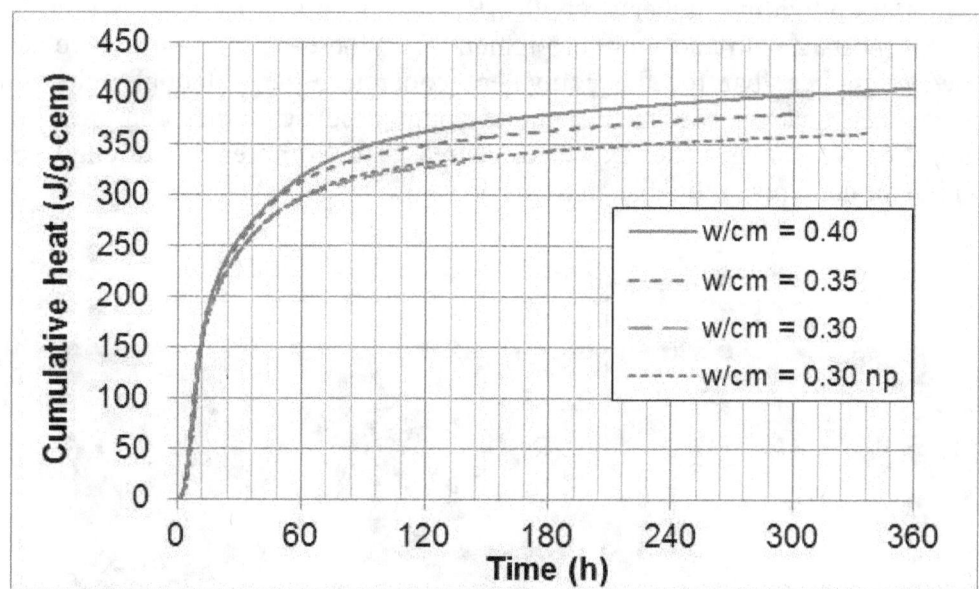

Figure 8. Cumulative heat release per mass of cement for the first 14 d of hydration for the four mortar mixtures.

Figure 9 shows results for mortars hydrated at two different temperatures (25 °C and 40 °C), to illustrate that the hydration of this blended cement can be accurately described using a commonly applied maturity-based approach, with a calculated apparent activation energy of 30.4 kJ/mol. The time-temperature transformed curves in Figure 9 nearly overlap one another, indicating the adequacy of the applied Arrhenius model.

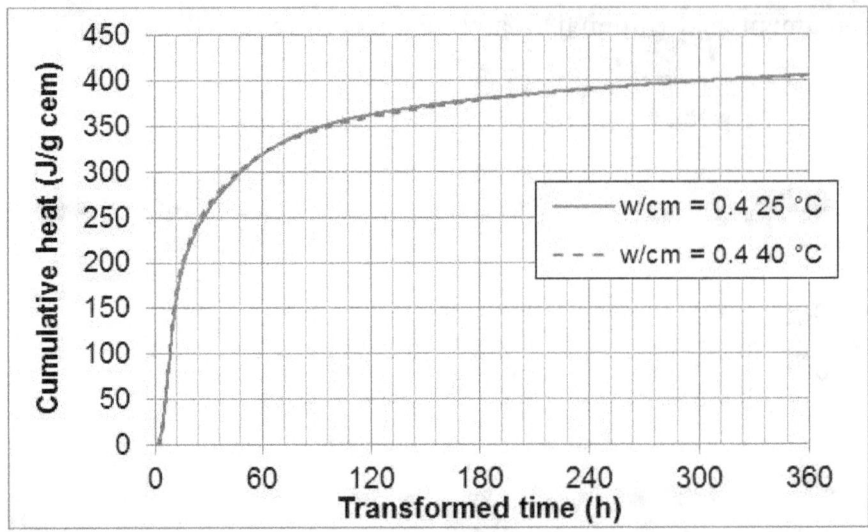

Figure 9. Example of time-temperature transformation, using an apparent activation energy of 30.4 kJ/mol, for *w/cm* = 0.40 mortar under isothermal conditions at either 25 °C or 40 °C.

3.2.4 Compressive Strengths

The measured mortar cube compressive strengths (average of three specimens) at ages of 1 d, 7 d, and 28 d are shown in the Figure 10. The trends are as would be expected with the lower *w/cm* mortars providing higher strengths. Relatively little difference is seen between the two *w/cm* = 0.30 mortars with and without pigment. The increase in compressive strength with decreasing *w/cm* implies that for the equivalent concretes, the potential reduction in *w/cm* between the batched and the specified concrete could produce higher compressive strengths, which could effectively mask the expected decrease in compressive strength due to any abnormally high air contents in the delivered product.

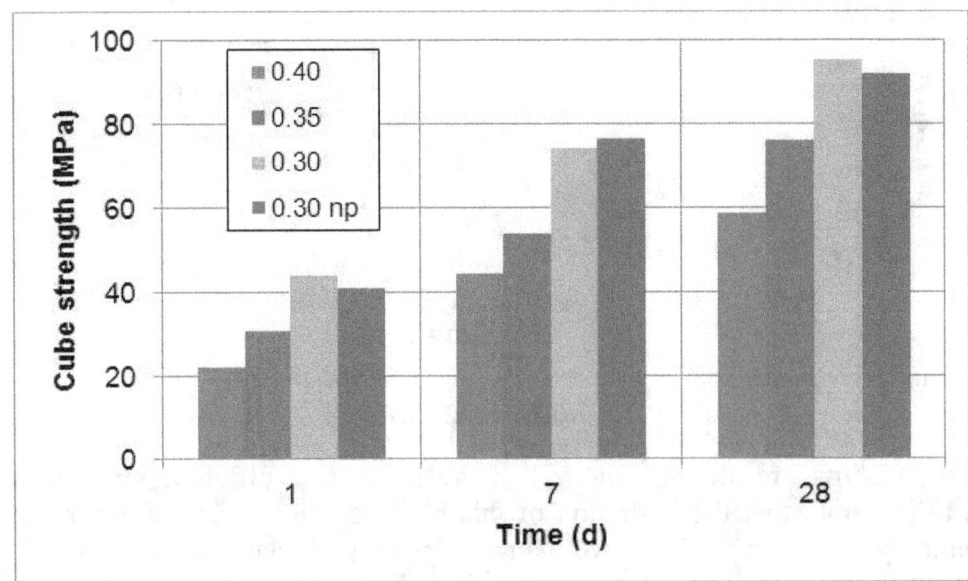

Figure 10. Measured mortar cube compressive strengths for the four mortar mixtures at ages of 1 d, 7 d, and 28 d. Mean coefficients of variation were 1.4 %, 1.7 %, 4.2 %, and 5.3 % for the 0.40, 0.35, 0.30, and 0.30 np mortars, respectively.

14

3.2.5 Autogenous Shrinkage

Autogenous shrinkage (ASTM C1698 [16]) measurements on the mortars were carried out to an age of up to 56 d as shown in Figure 11. It is apparent from this figure, that as is normally the case, a reduction in *w/cm* produces a mortar with measurably increased autogenous shrinkage. The measured shrinkage for the *w/cm* = 0.30 mortars at 28 d, for example, is about double that of the *w/cm* = 0.40 mortar. The potential decrease in *w/cm* of the delivered concrete (0.31 to 0.32) vs. that of the accepted concrete mixture (0.40) would significantly increase the autogenous shrinkage of the in-place concrete. Even a reduction in *w/cm* from 0.40 to 0.35 would significantly increase the autogenous deformation of the concrete. The influence of the pigment on the autogenous deformation of the *w/cm* = 0.30 mortar is well within the variation in the experimental measurement. It should be emphasized that for the concrete in the field, autogenous and drying shrinkage are always occurring simultaneously and it may be the combination of the two that ultimately leads to unwanted early-age cracking.

Autogenous shrinkage, as seen in Figure 11, is fairly linear over the entire measurement time scale of 56 d. Normally, for an ordinary Portland cement, the autogenous shrinkage increases rapidly and then levels off, following more of an exponential-shaped curve. For this particular blend of cement, as evidenced by the chemical shrinkage results, the pozzolanic reactions become very active at ages of about 7 d and beyond. It is thought that this pozzolanic reaction is responsible for maintaining a significant rate of autogenous shrinkage at ages beyond 7 d, and likely even beyond 56 d.

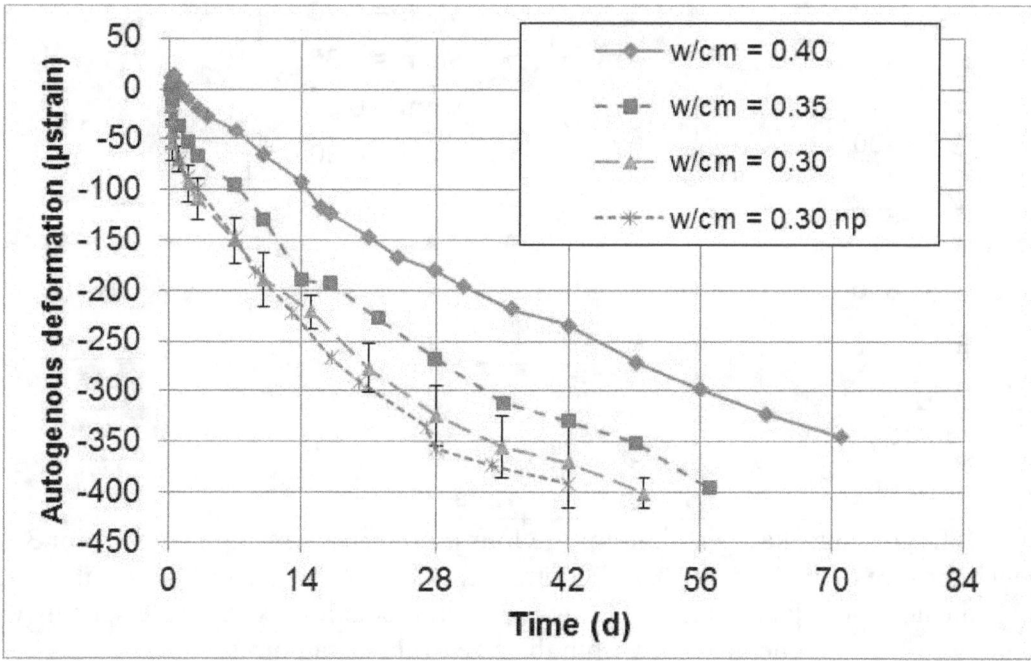

Figure 11. Measured autogenous deformation vs. time for the four mortar mixtures under sealed curing conditions. To provide an indication of measurement variability, error bars are shown for the *w/cm* = 0.30 mortar without pigment that represent ± one standard deviation based on the evaluation of three replicate specimens.

3.2.6 Drying Shrinkage

The same four mortar mixtures were evaluated for drying shrinkage based on the ASTM C596 standard test method [16]. The shrinkage tests were initially carried out in a controlled environment of 23 °C and 50 % RH for 56 d (environmental chamber). At that time, the specimens were moved to a second environmental chamber that was set at 23 °C and 25 % RH to evaluate further shrinkage. This was deemed prudent as the daily average relative humidity reported by the Overton, NV weather station in February and March of 2010 reached 25 % and even lower (Figure 3). Finally, the specimens were oven dried at 100 °C to determine the evaporable moisture content of each mixture. At each measurement time, both the mass and length of each of the six specimens per mixture were assessed. In addition, at ages out to 28 d, the resistivity of the specimens was also measured, using a commercially available 4-probe resistivity meter. The results are presented in Figures 12 to 16. Figure 12 simply shows the measured drying shrinkage (in microstrain) vs. time for the four different mortar mixtures. Unlike the autogenous shrinkage results, the measured drying shrinkage results for these four different w/cm mortars are quite similar. At later ages, about 10 % less drying shrinkage is observed in the w/cm = 0.30 mortars. Considering these drying shrinkage results along with the autogenous shrinkage results presented above (although the two cannot be simply added due to the 2 d of water curing applied in the ASTM C596 standard test procedure [16]), the total shrinkage should be generally greater in the mortars with the lower w/cm, likely increasing their propensity for early-age (total shrinkage) cracking.

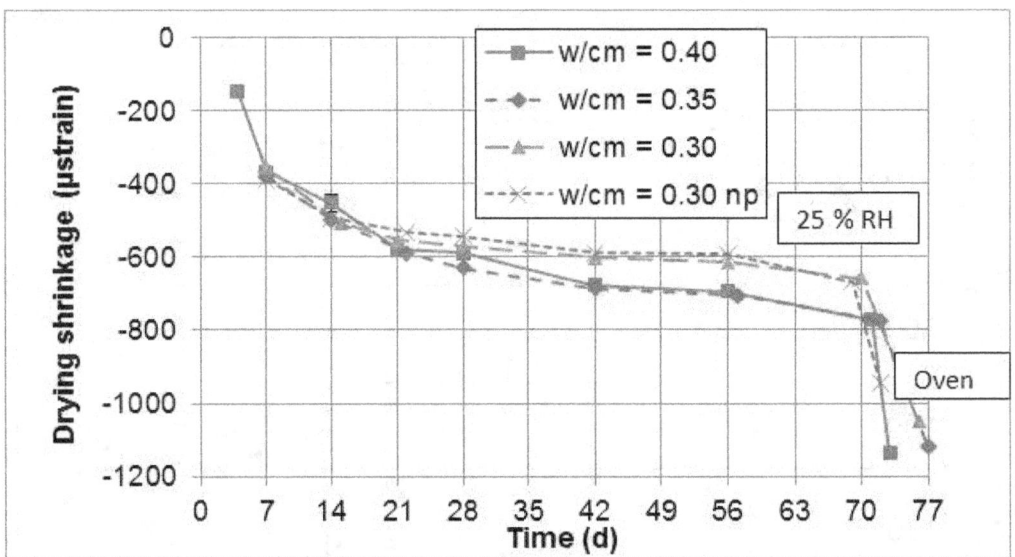

Figure 12. Drying shrinkage vs. time for the four mortar mixtures upon curing for 3 d and subsequent exposure to a 23 °C, 50 % RH controlled environment. Error bars for the *w/cm*=0.40 mortar indicate ± one standard deviation for measurements for six replicate specimens and generally fall within the size of the data points.

The ASTM C596 standard [16] suggests that the percent drying shrinkage be plotted against the reciprocal of the exposure time. These results are provided in Figure 13, which indicates the linearity expected when the data are plotted in this manner, with some minor deviations at early and late ages. Figure 14 plots measured shrinkage vs. measured mass loss (normalized by the initial demolded specimen mass) and a generally linear relationship is

16

obtained for each individual mixture. Figure 15 provides a plot of shrinkage vs. estimated mass of water remaining (using the final oven dry mass of the specimen and normalizing by the initial demolded specimen mass). The plot indicates that the data for the four different mortar mixtures converges towards a single line. This suggests that for these mortars prepared with the same starting materials, the measured shrinkage is controlled by the mass fraction (or volume fraction) of water remaining within the prisms as they are exposed to drying conditions.

Figure 16 contrasts the measured specimen resistivity to the measured drying shrinkage. The resistivity was measured by first dampening the four pins of the resistivity meter (spaced at an interval of 37 mm) and then centering them on the face of the drying specimen prism that was opposite to its top finished surface. Although there is some scatter, the plot indicates a potential master curve for the relationship between these two variables, perhaps suggesting that both are equally sensitive to the volume percentage and spatial arrangement of (liquid) water within the drying cement paste microstructure. The average coefficient of variation for the resistivity measurements for a single mortar mixture ranged between 3 % and 7.5 % for measurements performed between 3 d and 28 d. Conversely, for the shrinkage measurements, these coefficients of variation ranged between 1.5 % and 3.7 %.

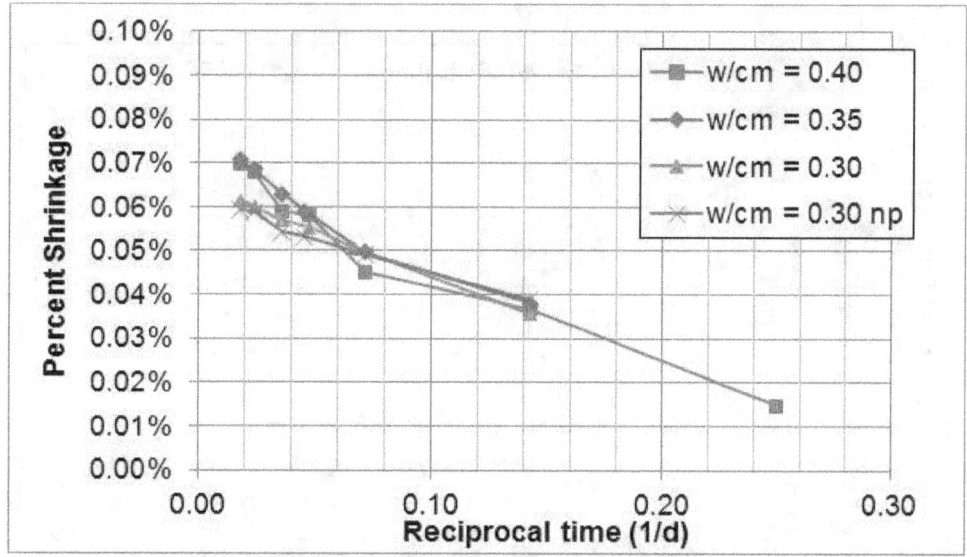

Figure 13. Measured percent shrinkage plotted vs. reciprocal time for the drying shrinkage data for the four mortar mixtures.

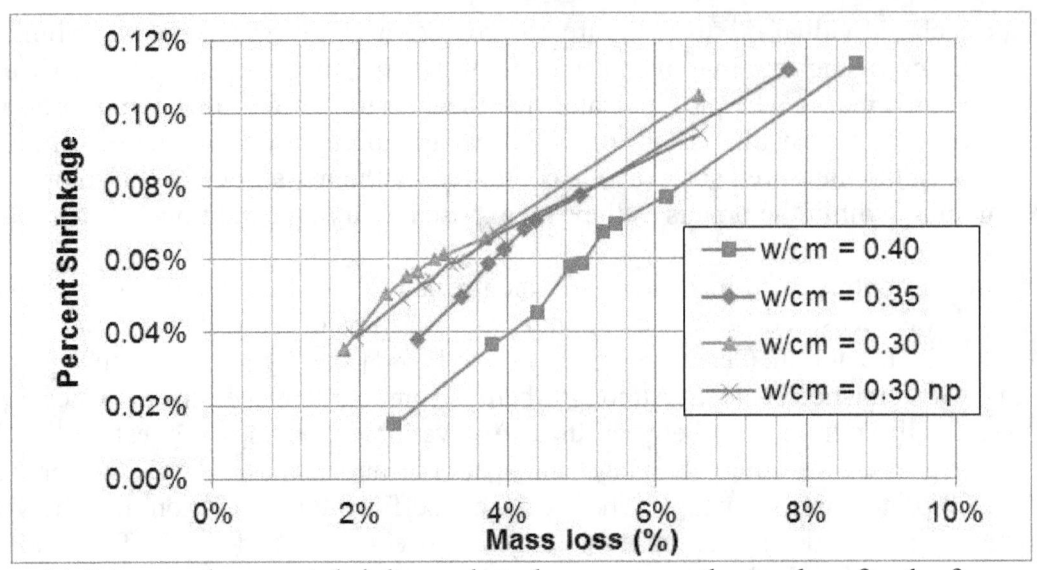

Figure 14. Measured percent shrinkage plotted vs. measured mass loss for the four mortar mixtures.

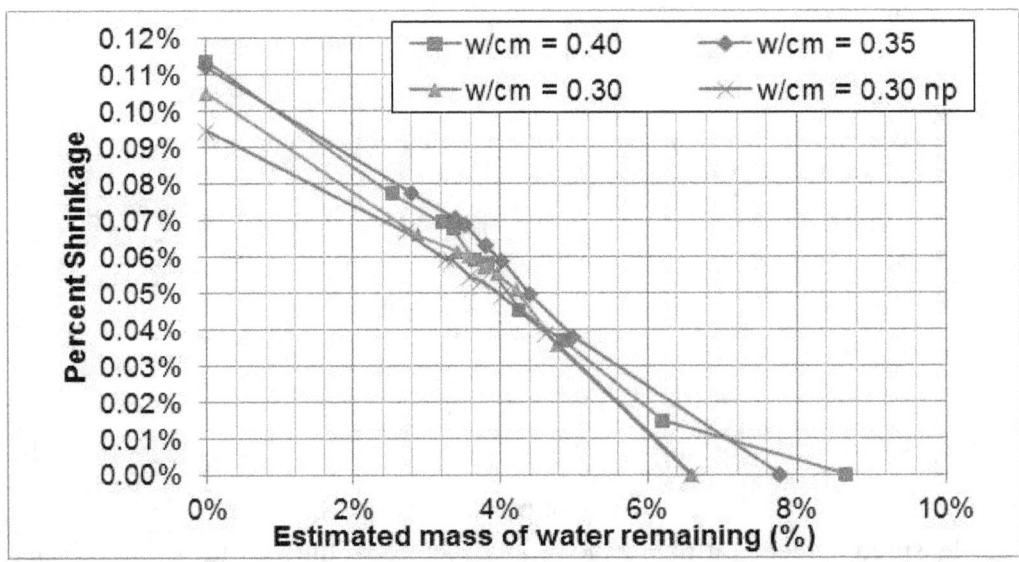

Figure 15. Measured percent shrinkage vs. estimated mass of water remaining for the four mortar mixtures.

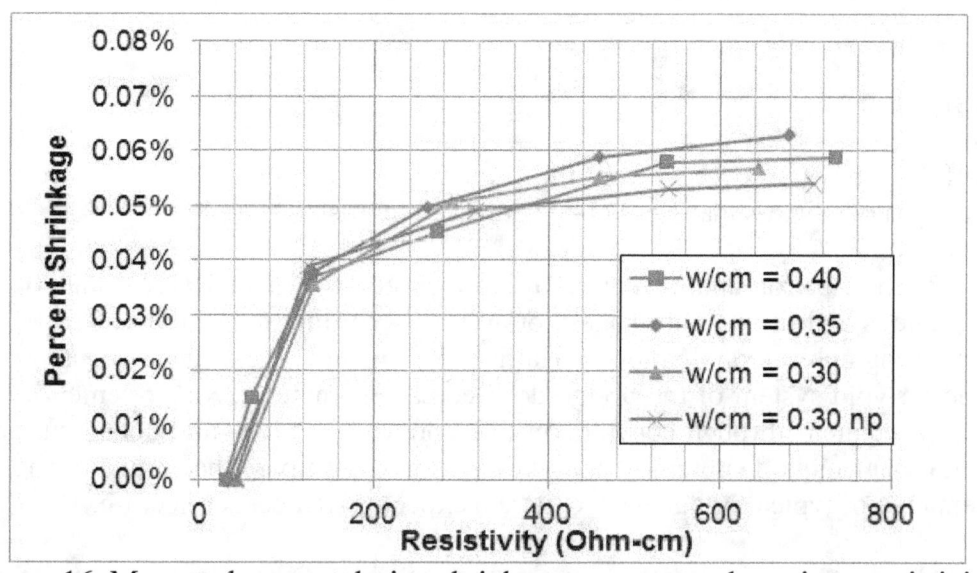

Figure 16. Measured percent drying shrinkage vs. measured specimen resistivity.

3.2.7 Restrained Ring Shrinkage

Depending on the mixture batch size, either one or two restrained ring shrinkage specimens were cast for each mortar mixture. Their times to cracking upon storage in the 23 °C, 50 % RH environment are provided in Table 4. The specimens cured under sealed conditions for 3 d experienced minimal mass loss, while those cured at 99 % RH still exhibited a measureable mass loss, typically on the order of 2 %. It was noted that the time to cracking, in addition to being influenced by *w/cm* was also likely influenced by "damage" present in a few of the specimens that were inherently more difficult to cast into their ring molds due to their less favorable rheological properties. This was observed in particular for the *w/cm* = 0.30 mortars with the pigment, where the difficulty of properly compacting the mortar into the ring molds likely induced some level of (pre)damage that reduced the subsequent time necessary for critical stress development and cracking to occur.

Table 4. Time to cracking observed for 25 mm thick ring specimens for each mortar mixture.

Mortar	3 d sealed curing	3 d 99 % RH curing
w/cm = 0.40	17 d	N.A.
w/cm = 0.35	42 d	> 56 d
w/cm = 0.30	7 d	7 d
w/cm = 0.30 no pigment	27 d	> 56 d

4. Tasks 2 and 3.

The following is an excerpt from the agreed upon statement of work between NIST and FHWA, specifying the items that were to be addressed in Tasks 2 and 3.

B. **Task 2** - Perform petrographic analysis of Northshore and Snake River bridge deck cores as well as autogenous and drying shrinkage mortar bars according to ASTM C 856-04, "Standard Practice for Petrographic Examination of Hardened Concrete" including a visual evaluation of the entrained air void system of the bridge deck cores. Estimate the water-cementitious ratio of the field core specimens through comparison to laboratory samples, and document and contrast cracking and other unique features by photomicrographs. Compare the results of the analysis to that representative of typical bridge deck concrete expected to meet service life.

C. **Task 3** – Conduct scanning electron microscopy (SEM) of the microstructure of Northshore and Snake River bridge deck cores, and the microstructure of the autogenous and drying shrinkage mortar specimens. Document and contrast the cracking and other unique features. Compare the results of the analysis to that representative of typical bridge deck concrete expected to meet service life.

4.1 Petrographic Observations on Bridge Deck Cores

Four sample bags containing cut drilled cores approximately 100 mm in diameter and 150 mm deep (partial depth) were delivered to NIST. In addition, four bags containing 50 mm partial-depth cores were provided by FHWA. Lapped sections were used for petrographic inspection using the stereo microscope and slabs were cut using a diamond blade wafering saw for subsequent scanning electron microscopy (SEM) analyses. These sections were the entire length of the available section. The phenolphthalein-stained sections were evaluated and a minimal carbonation depth was determined.

Echo Wash cores had a ground surface that produced grooves a few millimeters deep while the Valley of Fire cores had a broomed surface and with an epoxy coating, particularly with the core cut across a crack.

4.1.1 Echo Wash #1 and #2

Echo Wash core #1 was drilled on a crack that travels the length of the 135 mm section. The surface of the pavement is ground to produce grooves a few millimeters deep. The concrete appears strong, cutting and polishing well. The hardened cement paste bond to both the coarse and fine aggregate is strong as seen by a clean polish across the boundary and lack of aggregate plucking. The coarse aggregate was a crushed, angular dolostone and limestone to 20 mm size, and the fine aggregate was an angular crushed sand. The aggregate and paste distribution appears irregular, with numerous mortar-rich regions and coarse aggregate-rich regions. The coarse sand fraction appears to be minimal, leading to a gap-graded appearance. The cement paste appears strong and is subtranslucent, and tinted a pink-tan color with some of the pigment appearing to be partially absorbed into the coarse aggregate. Some evidence of retempering is apparent

20

against coarse aggregate, as a more dense, lower-air paste that appears darker. The entrained air void system appears to be abundant to excessive and entrapped air voids are also common. The crack in EW #1 fits together (unlike a plastic shrinkage crack that is of variable width), is roughly vertical through the core section and generally travels through the mortar. A few coarse aggregate were plucked out in the process of sectioning the slab.

4.1.2 Valley of Fire #1 and #2

The Valley of Fire cores are partial-depth cores of about 135 mm in depth and 100 mm in diameter and are from a pavement that has a brushed finish. Core #1 was drilled on a crack that extends the depth of the core. The concrete appears strong, cutting and polishing well. The hardened cement paste bond to both the coarse and fine aggregate is strong. The coarse aggregate is predominantly a crushed angular dolostone with some limestone, and the fine aggregate is also a crushed dolostone and limestone. The mortar and coarse aggregate is heterogeneously distributed with mortar-rich regions being common. The aggregate gradation also appears to be non-uniform with the coarse sand fraction missing. Abundant air is seen in these cores and some indication of a heterogeneous air distribution is seen with regions of abundant fine-sized bubbles, while other regions contain fewer, but larger air voids. Entrapped air is present, but appears less common than with the Echo Wash concrete. The hardened cement paste is subtranslucent and tinted a pink-tan color with some of the pigment absorbed into the coarse aggregate. The paste appears strong and polishes well. VOF #2 does have a crack that is parallel to the surface about 3 mm to 4 mm beneath the surface that intersects some entrapped air voids. The cracks in VOF #1 generally travel around the aggregate, with the exception of two large fractured aggregates.

Both sets of concrete cores are darker, or redder in appearance than the lab-produced mortar test cubes, suggesting a difference of pigmenting agent, or perhaps a lower *w/cm* ratio, making the concrete darker in color.

SEM analysis of a sample taken from each coring location was performed to evaluate the condition of the hardened cement paste, the residual cement, and the air void system. Mortar cubes with *w/cm* bracketing that specified for the concrete were also evaluated to compare the paste microstructures at different water contents to the concretes, to see if this might be a means to assess the true in-place concrete *w/cm* ratio. Imaging of the paste fraction and attempts to reconstruct the original cement grain distribution based upon the relatively immobile magnesium content failed due to the low MgO content of the cement and lack of Mg in any significant level in the silicates.

Point count analyses were performed on the #2 cores from both locations to measure the volume fractions of the aggregate, hardened cement paste, and air (both entrained and entrapped) to develop an estimate for the water/solids ratio based upon calculated volumes from the provided mixture proportions.

Some attempts to color-correct micrographs have been made to better represent the actual tones of the concrete. The concrete slabs were imaged under natural, indirect light providing the best reproduction, while the light microscopy utilizes a source that tends to be more yellow. The cores from both locations have a similar hue and color saturation that is not consistently reproduced in the images here, but is an important feature in the evaluation of relative differences in porosity and pigmentation. The microscope image colors are somewhat less important in this instance as they are consistent among samples and seek to illustrate the textures of the hardened cement paste, the residual cement, and the air void system.

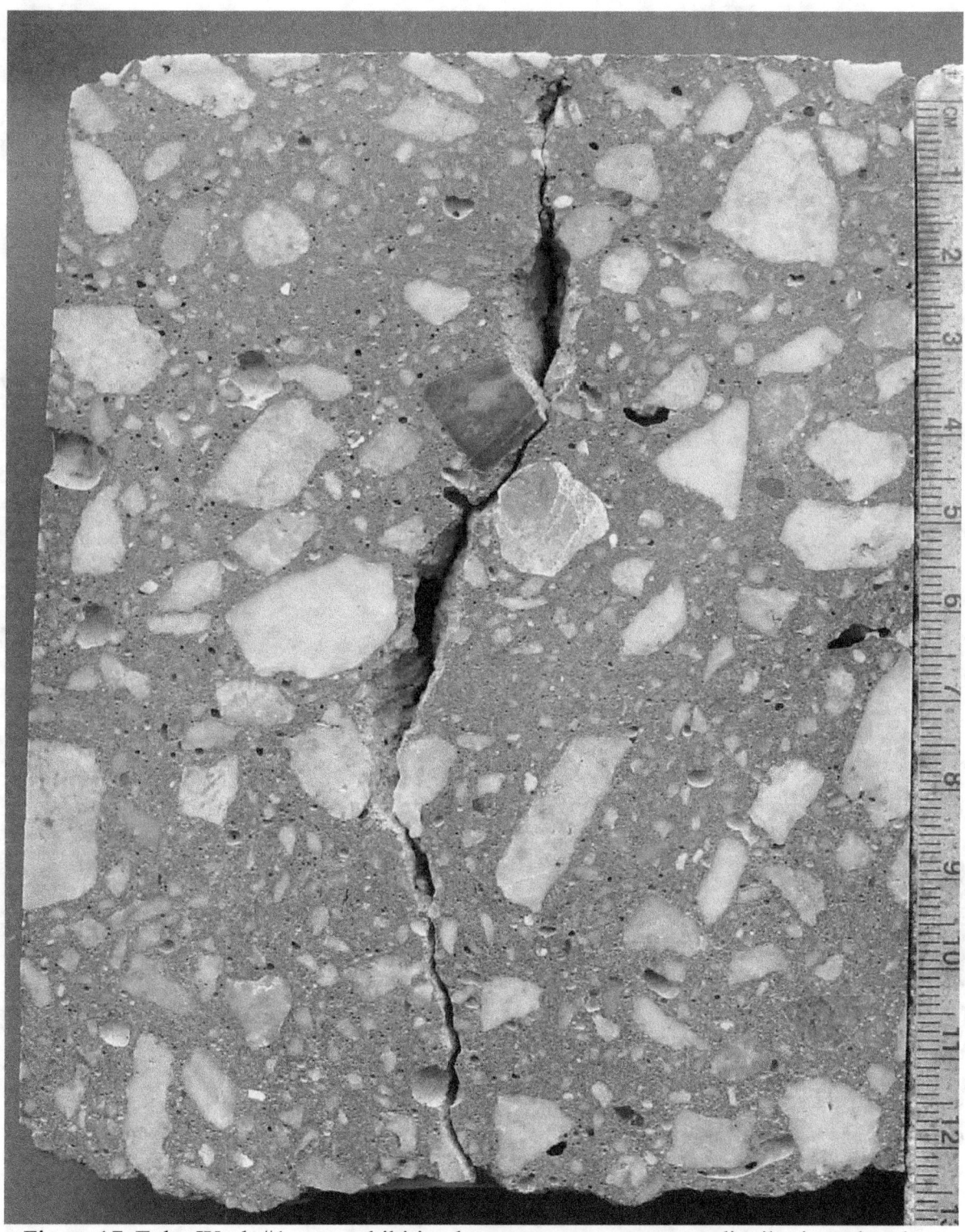

Figure 17. Echo Wash #1 core exhibiting heterogeneous aggregate distribution, abundant entrapped air and a crack extending the length of the core section.

Figure 18. Echo Wash #2 core exhibits abundant entrapped air and some heterogeneous aggregate distribution. The grooves from the surface cutting operation are visible at the top of the core.

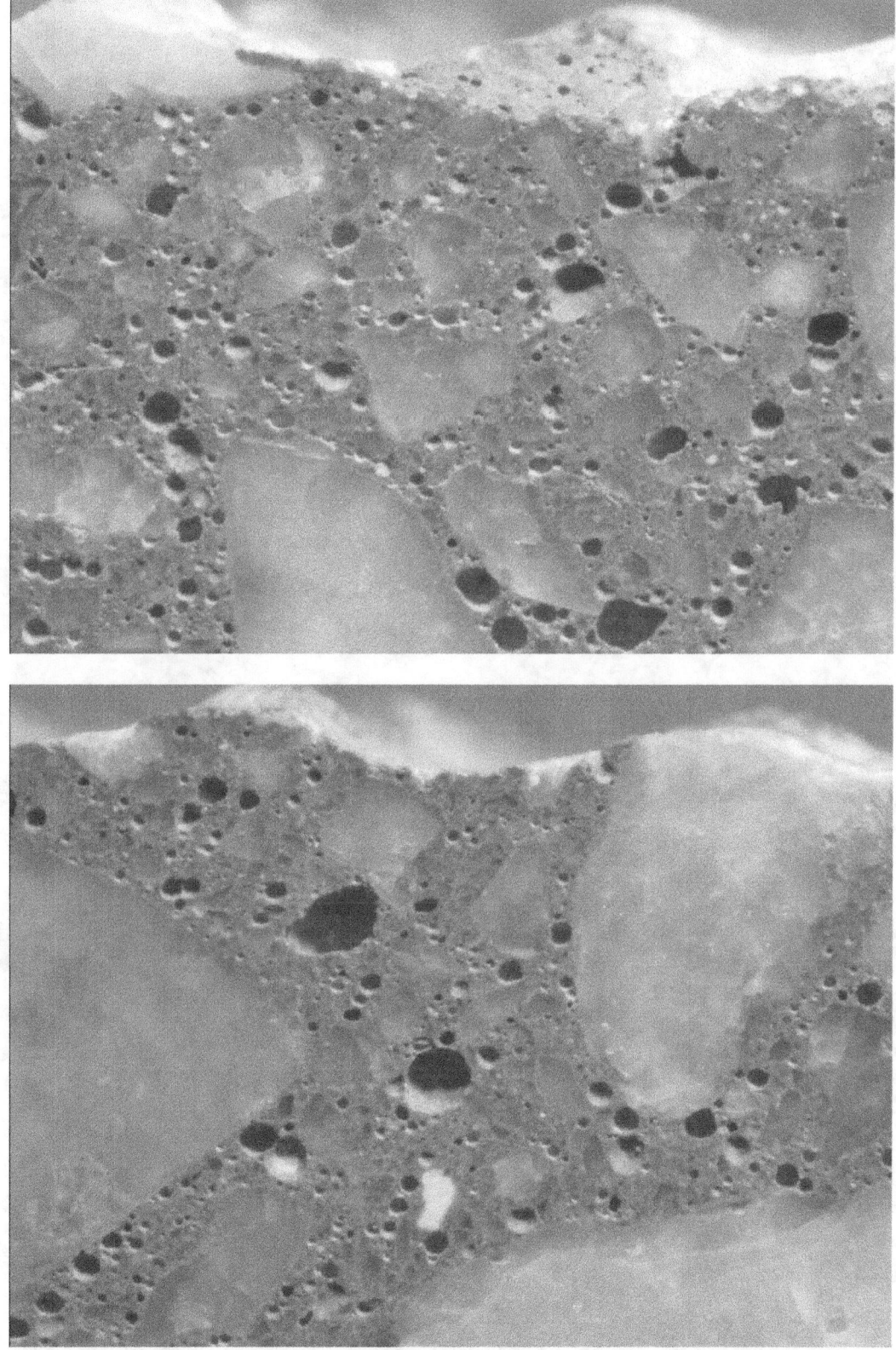

Figure 19. Surface microstructure of the Echo Wash #2 core with a field width of 10 mm shows the strong hardened cement paste by well-defined surfaces and air void edges, smooth cement paste, and the abundance of air.

Figure 20. Darker, relatively air-free paste adjacent to the coarse aggregate is present in some cases, suggesting retempering (upper image, 10 mm field width) while large entrapped air voids (lower image) are common throughout the core.

Figure 21. Comparison of the Echo Wash cores #1 (upper) and #2 (lower) against 0.30 (left) 0.35 (center) and 0.40 (right) *w/cm* 50 mm cubes of lab mortar shows that the deck cores are redder in appearance, suggesting a lower *w/cm* ratio or possibly an increased dosage of pigment.

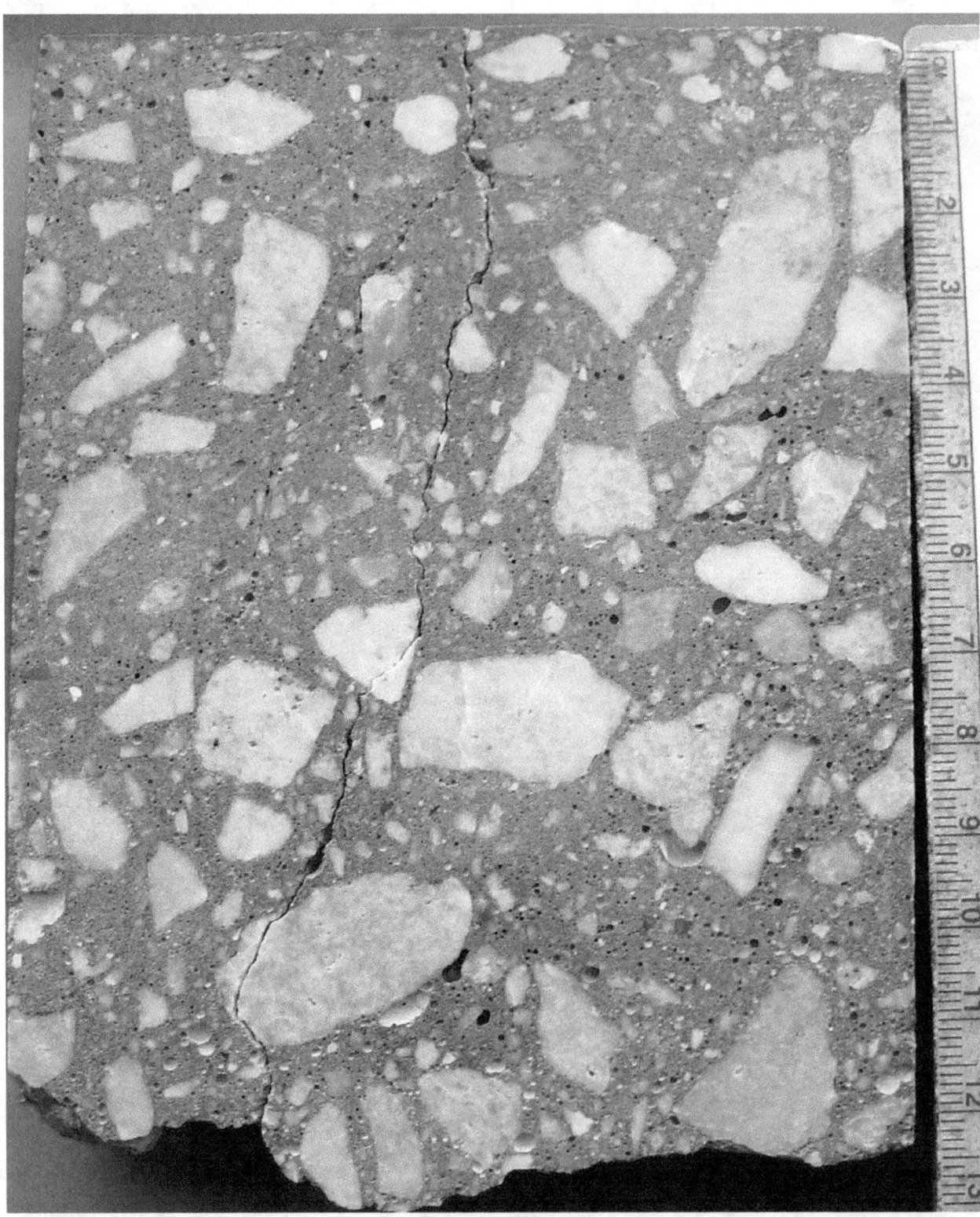

Figure 22. Valley of Fire core #1 with crack extending the length of the section passing mostly through the mortar with a heterogeneous aggregate distribution, common entrapped air, and heterogeneous entrained air void distribution.

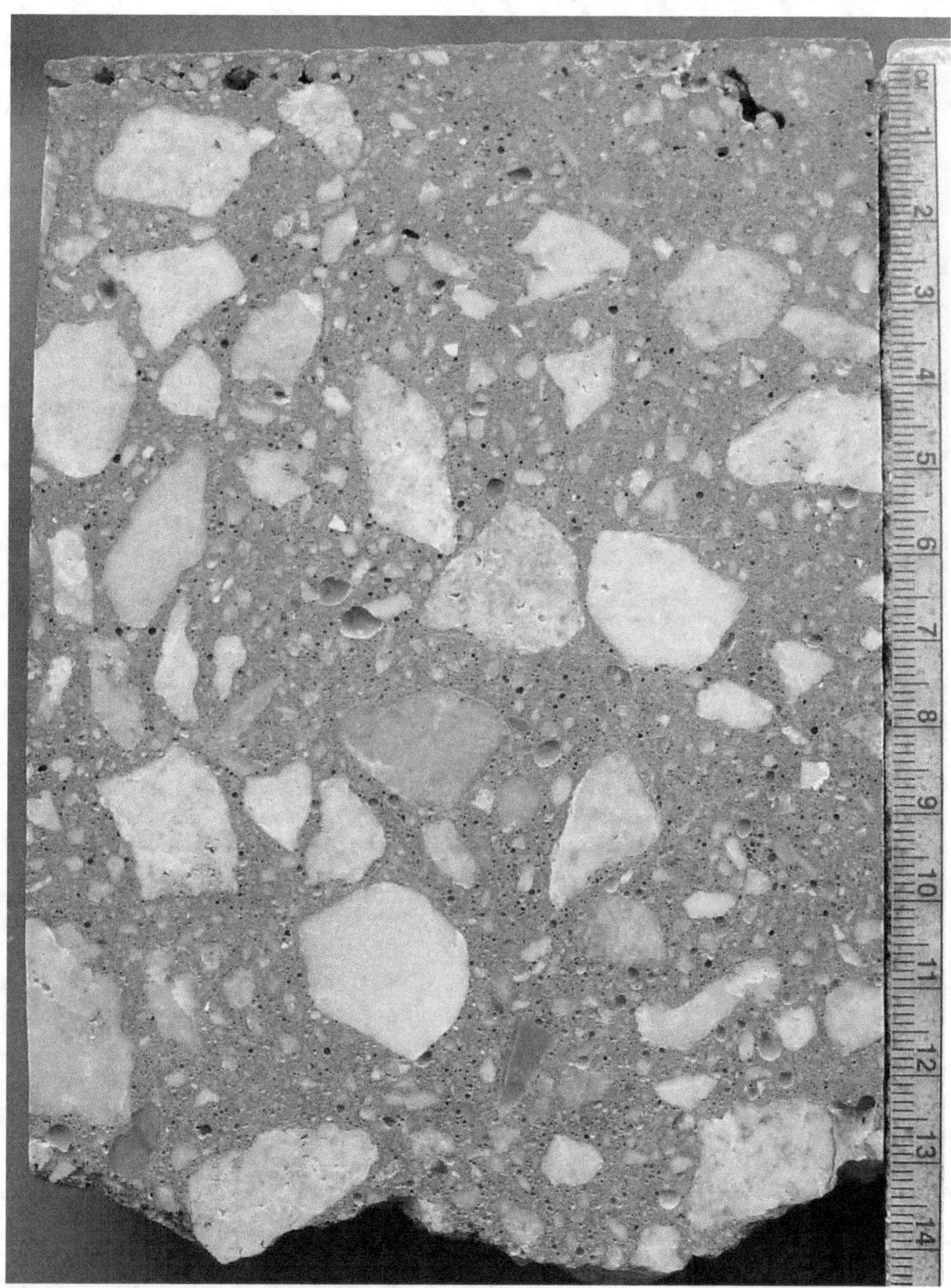

Figure 23. Valley of Fire core #2 with heterogeneous aggregate distribution, abundant entrapped air and crack parallel to the surface on the upper-left that transects entrapped air voids.

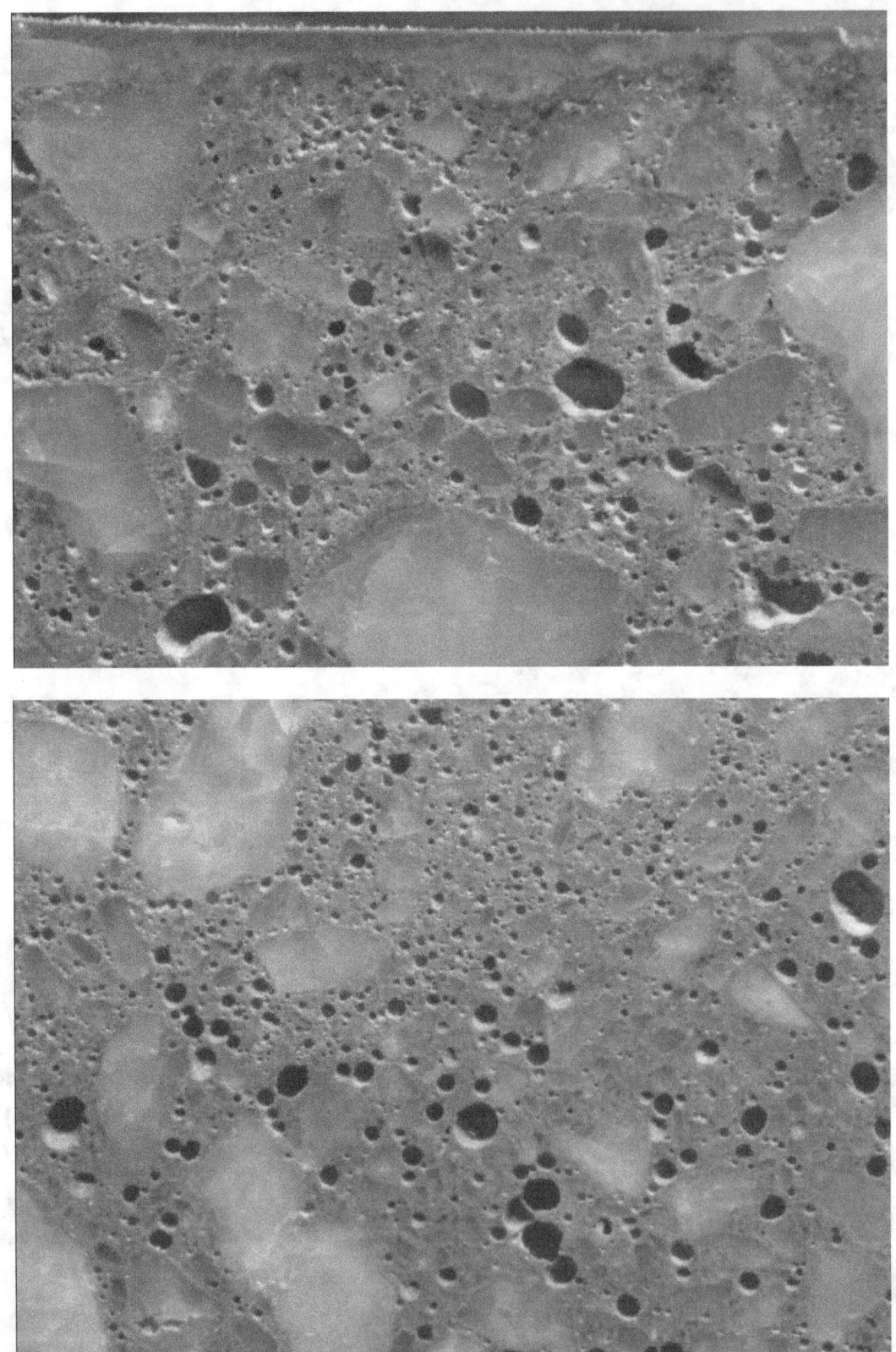

Figure 24. Valley of Fire core #1 surface (upper, 8 mm field width) showing the epoxy layer and permeation of a few mm into the core, abundant but variable air, and (bottom, 14 mm field width) image of numerous fine-sized (upper left of image) and fewer large-sized air voids.

Figure 25. Comparison of the Valley of Fire cores #1 (upper) and #2 (lower) against 0.30 (left) 0.35 (center) and 0.40 (right) *w/cm* 50 mm cubes of lab mortar shows that the deck cores are redder in appearance, suggesting a lower *w/cm* ratio or possibly an increased dosage of pigment.

4.1.3 Snake River Bridge Cores

Three cores from the set of Snake River specimens were examined, one approximately 110 mm in diameter by 110 mm deep drilled to just below an epoxy-coated rebar, and two 60 mm diameter by 120 mm deep cores labeled as #3 and #4. The surfaces of the cores are grooved and the finished concrete surface appears rougher than that of the Valley of Fire cores. These cores have high entrained air contents and fairly high entrapped air contents. The cement paste from this location appeared to be sound, well hydrated, and dense.

Each core was cut in half lengthwise to provide a surface for examination. That surface was lapped using a series of grits to produce a surface suitable for viewing under the stereo microscope and for point count analysis of the aggregate, paste, and air components. The microstructures of the concrete cores were similar, so a single general petrographic set of observations will be presented (Figures 26 to 28). The point count analysis provides an estimate of the aggregate, hardened cement paste, and air void area fractions and a means to compare the observed components with those calculated given the mixture proportions. The Snake River deck concrete cement paste appears strong, and it cuts and polishes easily. The cement paste is a uniformly gray color for all cores with no dense paste regions adjacent to the aggregate that are typical of retempered concrete. The edges of the paste along air voids remain sharp and the bond of the cement paste to the aggregate is strong. The coarse aggregate was a rounded siliceous gravel to about 20 mm in diameter and the fine aggregate was crushed silica sand. The aggregate distribution appears fairly uniform, with some mortar-rich regions observed in core #4. Each core contained a crack that extended the full core depth for #3 and #4, and 75 mm depth to the rebar for the large diameter core. The large core was treated with phenolphthalein solution to estimate the depth of carbonation, which is restricted to the upper-most mm layer of cement paste.

Figure 26. Snake River large-diameter core cross section has a crack extending from one of the surface grooves to the top of the epoxy-coated rebar. Some entrapped air voids are visible.

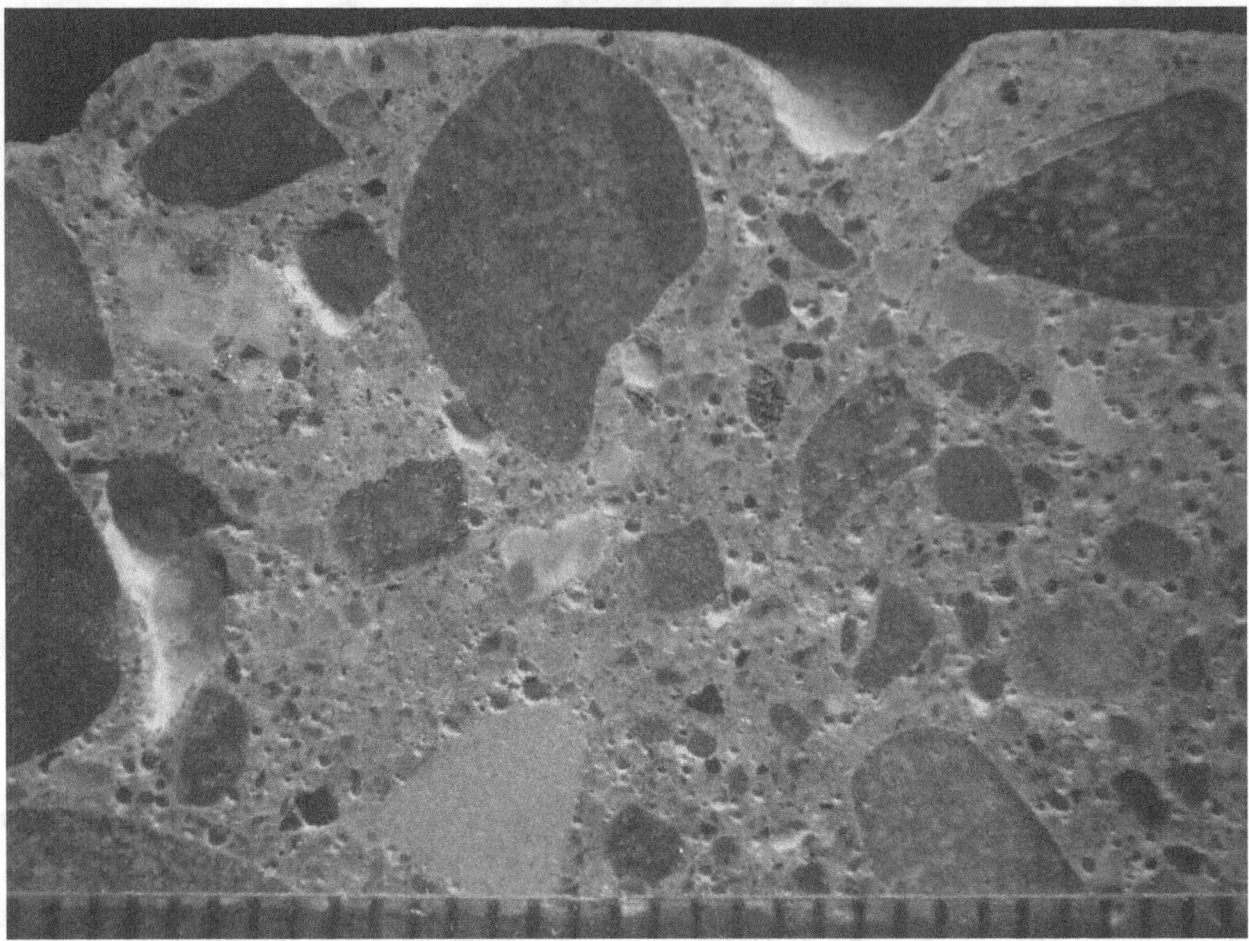

Figure 27. Snake River large core surface showing cut grooves, uniform aggregate distribution, entrained and some large entrapped air voids. Field width = 31 mm.

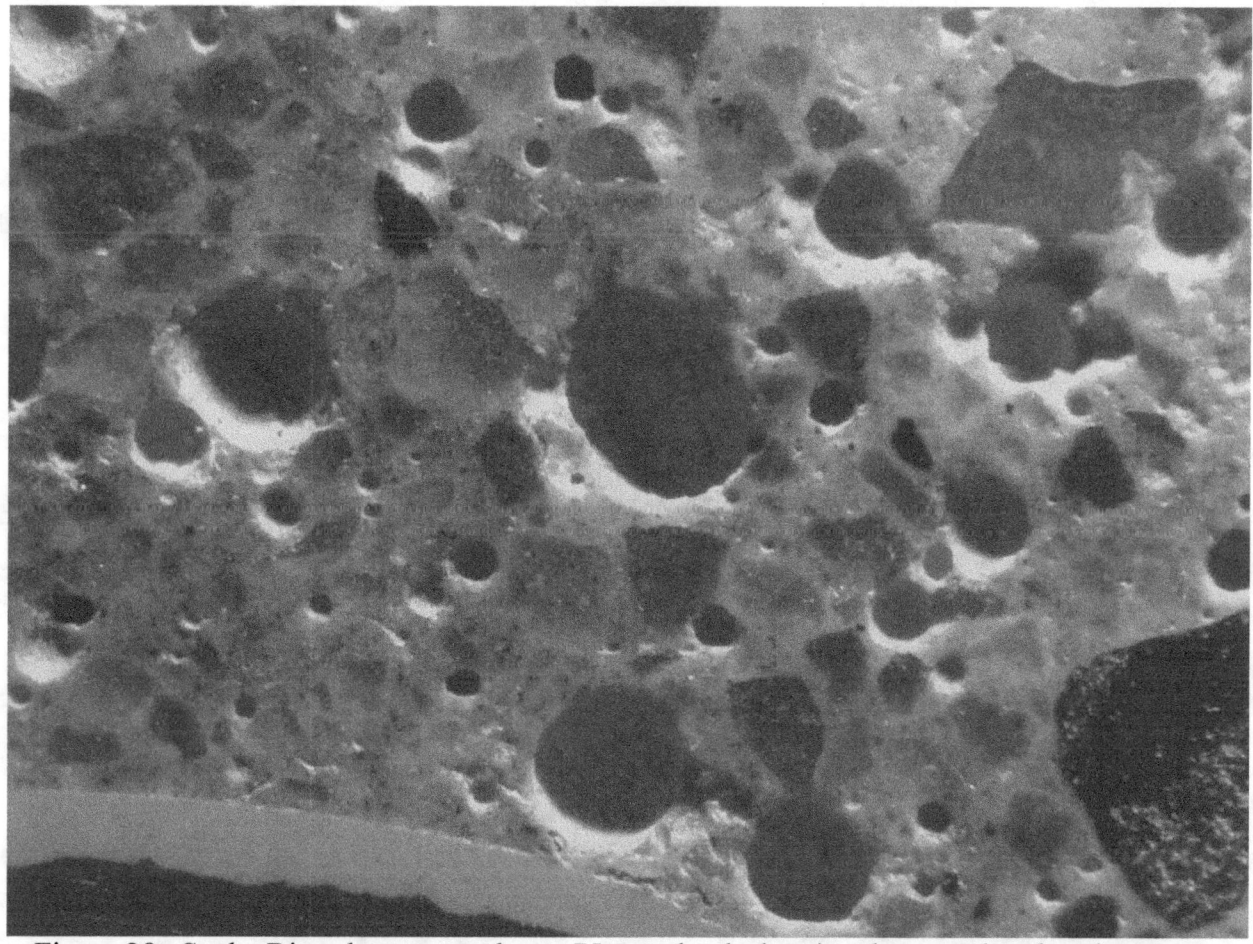

Figure 28. Snake River large core about -75 mm depth showing the paste bond to the epoxy-coated rebar and the entrained air void system.

4.2 Quantitative Light Microscopy

4.2.1 Echo Wash and Valley of Fire Cores

The air content was measured using a section taken from both core locations by carefully lapping the section, polishing the surface to better define the features, filling the voids with calcium hydroxide powder to provide contrast with the surrounding paste and aggregate, and video imaging to measure the area fraction of air. No distinction was attempted between entrained and entrapped air, and voids within the aggregate were removed prior to measuring the area fractions. The results for total air for each deck from core two may be found in Table 5. Variability by field may be attributed to the presence or lack of coarse aggregate.

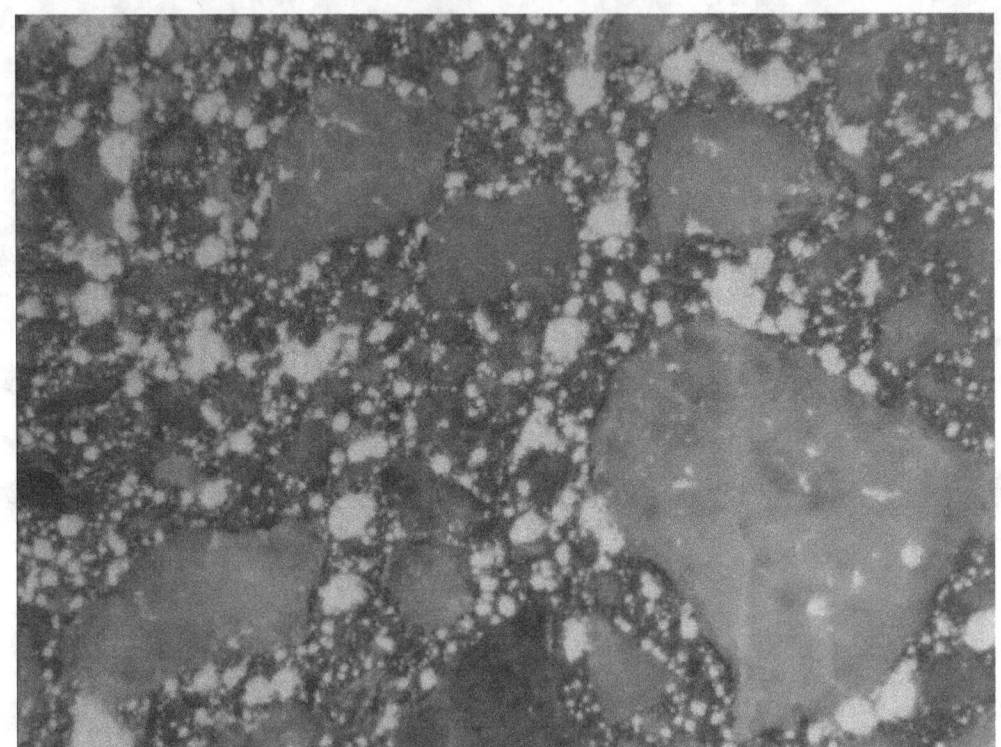

Figure 29. Calcium hydroxide powder filling air voids provides contrast necessary for quantitative video imaging. Field width: 25 mm.

Table 5. Area fraction of air voids from Valley of Fire and Echo Wash bridge decks using light microscopy and 25 mm image fields.

Field	Valley Of Fire	Echo Wash
1	12.98 %	14.05 %
2	10.07 %	7.67 %
3	15.36 %	9.93 %
4	10.64 %	13.79 %
5		11.25 %
Mean	12.3 %	11.3 %
1s	2.4 %	2.7 %

Point counting was used to establish estimates of the volume fractions of each of the components of each of the hardened concretes, namely the aggregate, the hardened cement paste, and the air voids. Through comparison of the volume fractions of the concrete constituents to that of the concrete mixture proportions with different amounts of water, it may be possible to estimate the concrete *w/cm* ratio. The point count process employs a regular grid of points and a light microscope where the magnification is set such as to balance the need to resolve finer-sized constituents, yet have reasonable spacing between points. The relationship between area fraction and volume fraction was established in the late 1800's and point count analyses may be found in ASTM C457 for air voids and ASTM C1356 [16], a microscopy-based procedure to assess the area fraction of clinker phases within the construction materials standards. The results are typically expressed on a volume basis (but can be expressed on a mass basis after correcting for the true densities of each phase), and the uncertainty of the estimate of an analysis is related to the total number of points counted, with the absolute uncertainty at the 96 % confidence interval given by:

$$\delta = 2.0235 \sqrt{\frac{P(100-P)}{N}} \tag{1}$$

where

δ = the absolute error in percent,
P = the percentage of points of a phase, and
N = the total of all points.

Core #2 from both bridge decks were subjected to a point count analysis using a field width of about 5 mm selected as a balance between resolution and sampling rate. A grid 6 points by 7 points was superimposed on the image and the phase falling under each point – aggregate, cement paste, or air, was recorded. The coarse and fine aggregate was counted as a single category and no distinction was attempted for entrained and entrapped air. On an air-free basis, these results may be compared to calculated volume fractions of aggregate and paste based upon *w/cm* ratios of 0.40 and 0.32 (Table 6, right columns) and best conform to the 0.4 *w/cm* proportions within the uncertainty of the measurements. In fact, this analysis suggests that the *w/cm* proportions of the in-place concrete may have been slightly above 0.40, particularly for the Valley of Fire concrete.

Table 6. Point count results from single cores extracted from the Echo Wash and Valley of Fire bridge decks expressed as area fraction, 96 % uncertainty confidence level and total points counted.

Echo Wash	Mean	Uncertainty	Points		Air-Free	*w/cm 0.4*	*w/cm 0.32*
Aggregate	58.6 %	1.3 %	3564		66.4 %	67.2 %	69.7 %
Paste	29.6 %	1.2 %	1800		33.6 %	32.7 %	30.3 %
Air	11.9 %	0.8 %	723				
Valley of Fire							
Aggregate	54.4 %	1.2 %	3871		65.1 %	67.2 %	69.7 %
Paste	29.1 %	1.1 %	2071		34.9 %	32.7 %	30.3 %
Air	16.5 %	0.9 %	1177				

4.2.2 Snake River Bridge Cores

Calculations of material volumes based upon the mixture proportions are provided in Table 7, while the results of the point count analyses, expressed as volume fractions with a 96 % uncertainty on each estimate are provided in Table 8. A total of 7250 points were counted on the large core, while cores #3 and #4 had totals of 2520 points and 2748 points, respectively. Entrapped air was estimated on the Snake River cores by separately tabulating voids in excess of 1 mm. This allowed a second calculation to estimate the entrapped air content. Volume fractions estimated for the large core and core #3 are consistent with the mix design. Core #4 volume fractions of paste and aggregate are different, which may reflect the relatively large mortar-rich area in that core. The air content for all cores is higher than the specified mix design air content, and for cores #3 and #4, it is substantially higher.

Images of cores #3 and #4 are presented in Figures 30 and 31, showing the lapped cross section from each core. A crack traverses the full depth of the core, but the sections did not separate after cutting and lapping. While visible under the stereo microscope, the crack width is fine and consistent from top to bottom while typically passing through the mortar fraction of the concrete. Entrapped air voids are common, are irregular in shape and can range up to 20 mm in width.

Higher magnification views of core #3 (Figure 32) show the entrained air void system and what appears to be some clustering, which can be interpreted as a sign of retempering. This clustering is generally confined to the mortar and not adjacent to coarse aggregate and no evidence of a denser, lower *w/cm* paste adjacent to the aggregate was detected. The point count analysis indicates that core #3 has an air content of 13.1 % with a 1.4 % uncertainty (96 % confidence interval) and the entrapped air content (included in the entrained air estimate) of 2.7 % (0.7 % uncertainty) is slightly high (Table 8). Higher magnification images of core #4 (Figure 33) show the irregularly-shaped entrained air voids and some of the large entrapped air voids. A comparison of core #3 to the large Snake River core (Figure 34) provides a visual confirmation on the relative differences in air content between these cores.

Table 7. Volume fraction estimates for the Snake River Bridge concrete mixture design and calculated air-free paste and aggregate volume fractions for a specified 0.39 *w/cm* concrete.

	Specific gravity	Mass (kg/m³)	Volume	V fraction	Air-Free	
		Snake River Concrete Mix Design				
Cement	3.15	312	99.05	0.099		
Fly ash	2.3	78	33.91	0.034		
Coarse aggregate	2.643	1015	384.03	0.385		
Fine aggregate	2.619	737	281.41	0.282		
Water	1	153	153.00	0.154	0.301	paste
0.39 *w/cm*			951.40	0.955	0.699	aggregate
		air content	0.05	1.005	1	

38

Table 8. Volume fractions and 96 % confidence interval uncertainty based upon point count analyses of three Snake River concrete cores.

	Volume	Uncertainty	Air-Free	Mix Design
SR Lg. Core				
Aggregate	63.7 %	1.1 %	0.696	0.699
Paste	27.8 %	1.1 %	0.304	0.301
Air	8.6 %	0.7 %		
Entrapped Air	1.2 %	0.3 %		
SR #3				
Aggregate	60.1 %	2.0 %	0.692	
Paste	26.7 %	1.8 %	0.308	
Air	13.1 %	1.4 %		
Entrapped Air	2.7 %	0.7 %		
SR #4				
Aggregate	51.5 %	2.0 %	0.588	
Paste	36.0 %	1.9 %	0.412	
Air	12.6 %	1.3 %		
Entrapped Air	2.7 %	0.6 %		

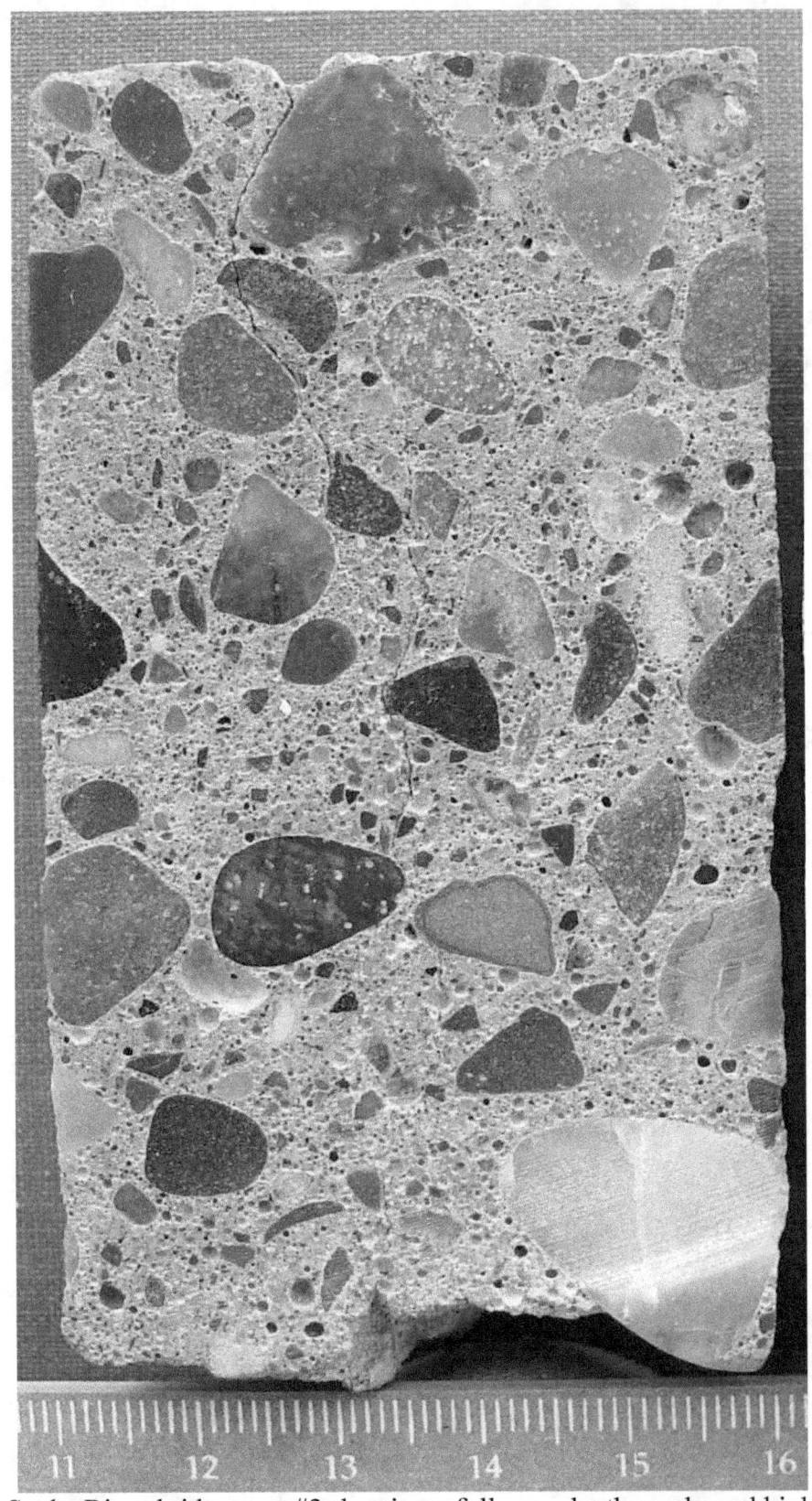

Figure 30. Snake River bridge core #3 showing a full-core depth crack, and high air content.

40

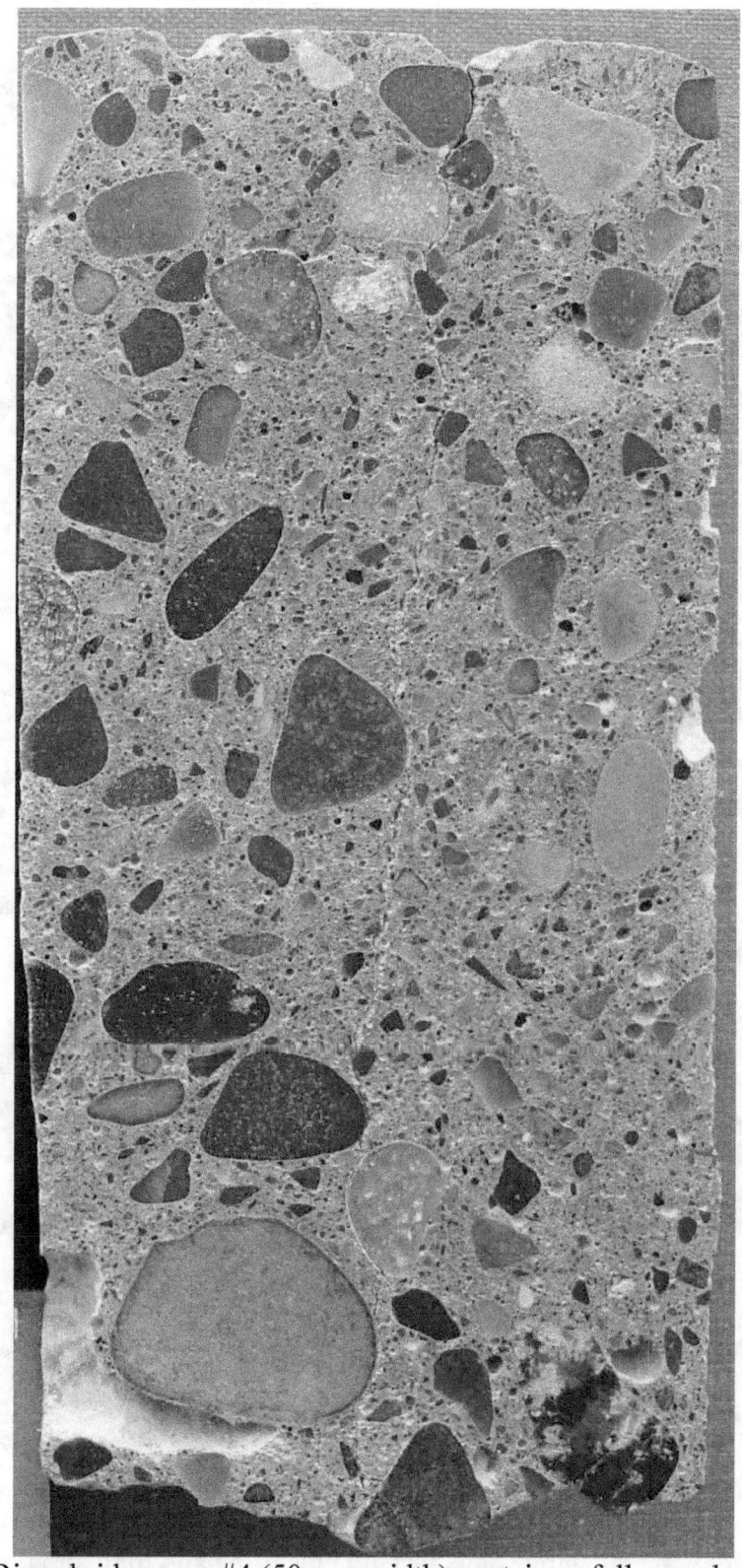

Figure 31. Snake River bridge core #4 (50 mm width) contains a full-core depth crack and some large entrapped air voids, mortar-rich regions and abundant entrained air.

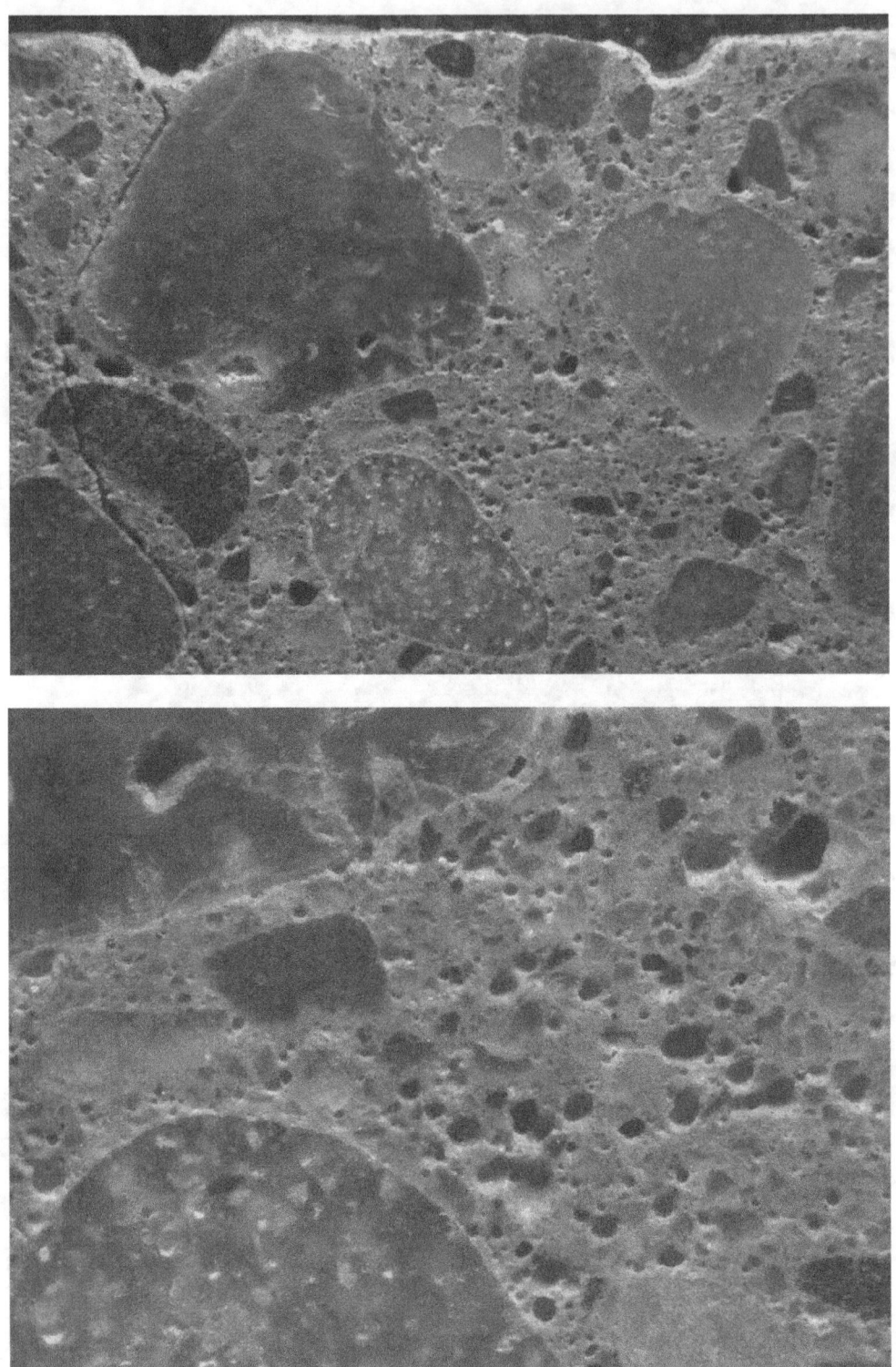

Figure 32. Snake River bridge core #3 showing the surface (32 mm field width) and crack traveling into the core (upper image) and a closer view (10 mm field width) of the hardened cement paste showing some air void clustering and high air content.

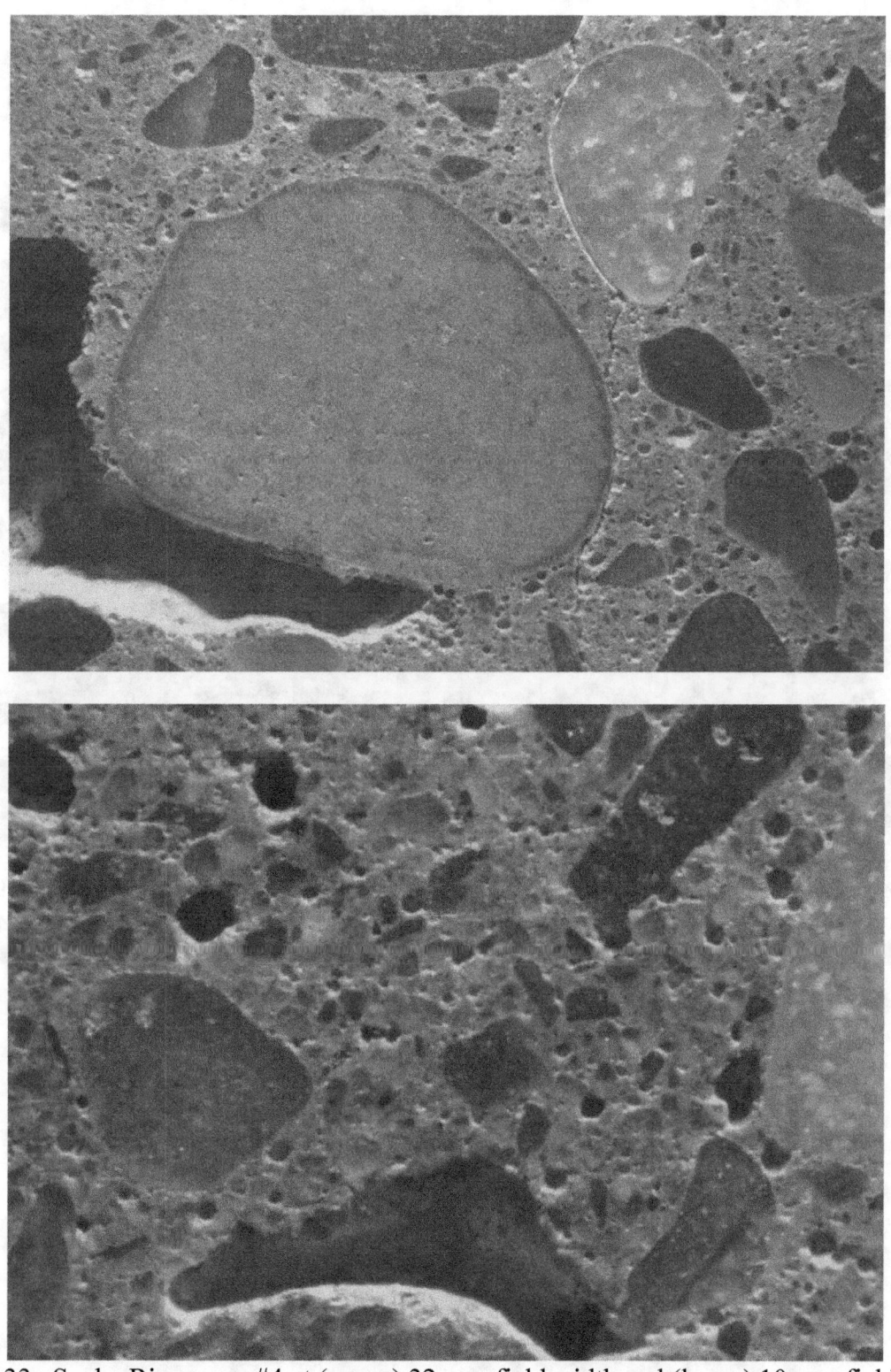

Figure 33. Snake River core #4 at (upper) 32 mm field width and (lower) 10 mm field width showing large entrapped air voids and high air content.

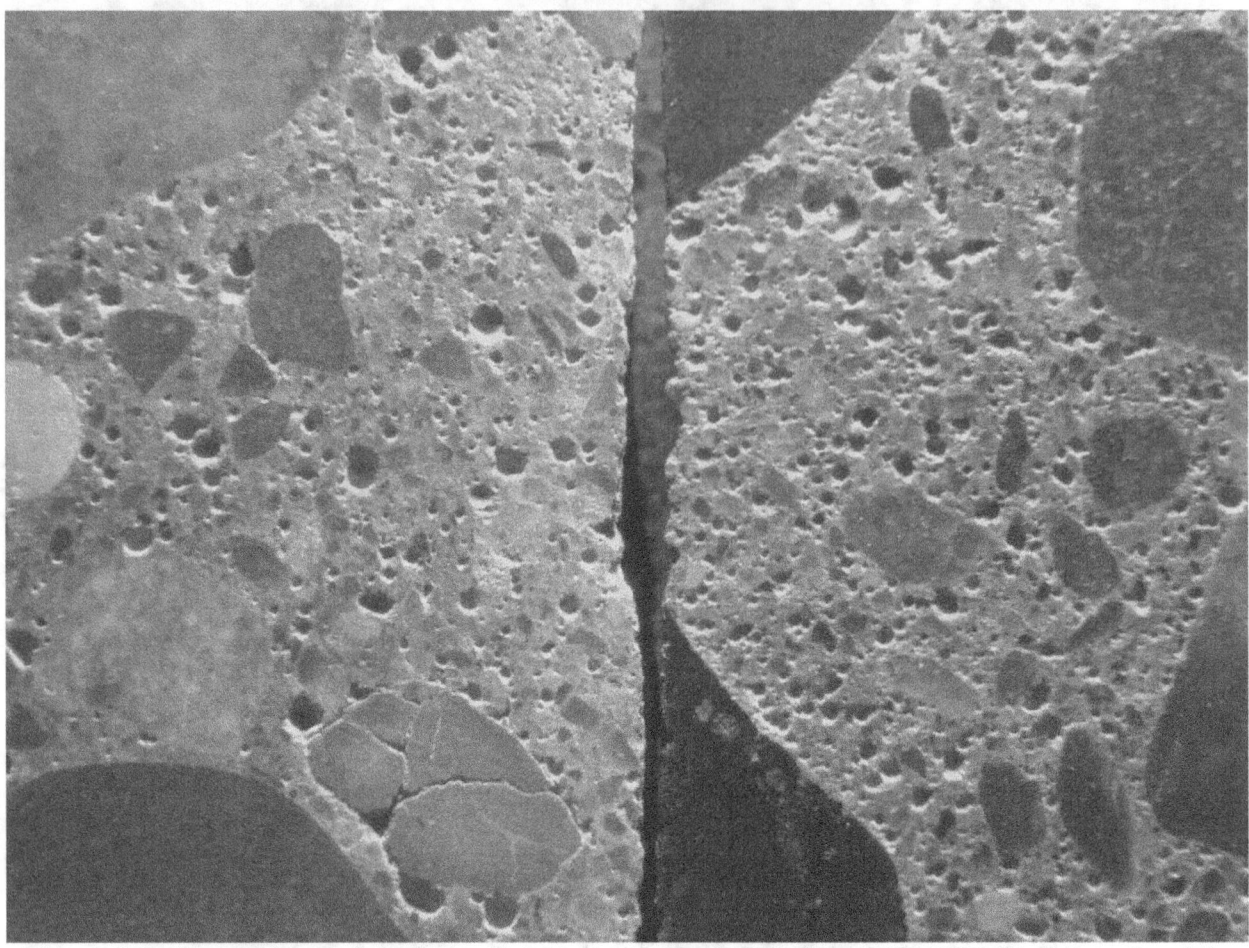

Figure 34. Comparison of the large (8.6 % air) on the left and core #3 (13.1 % air) on the right clearly shows the difference in air void volume and possibly a larger overall air void size for the large core (field width 32 mm).

4.3 Scanning Electron Microscopy

Backscattered electron (BSE) and X-ray images from the SEM provide another view of the concrete microstructure, with some distinct advantages. With the hardened cement paste being sub-translucent to the light microscope, contrast between cement hydration products is low and that with the remnant cement particles is also low, with the exception of the dark-colored ferrite phase. The SEM backscattered electron image contrast reflects the average atomic number of the phase, a function of the chemical composition and inherent porosity. Higher average atomic number phases will appear brighter, while lower ones will appear darker. X-ray microanalysis provides chemical information from a spot or region with the spectrometer measuring total counts at each x-ray energy channel. By establishing a window across the characteristic X-ray windows, a map of element spatial distribution may be generated, where the brightness is proportional to the element concentration. The spatial resolution of the BSE image is roughly 0.25 µm, and that of the X-ray image is roughly 1 µm.

SEM imaging was used to estimate the amount of residual cement and distribution of the concrete constituents and was used to image the paste fractions of each core location for comparison with mortar cubes prepared in-house to see if the w/cm affected cement paste texture and if we could qualitatively or quantitatively relate the core microstructure to one of the test cubes with known w/cm. A difficulty with this approach was in achieving an adequate sampling area that was sufficient to have a representative sample, so while the results are presented here, the point count results presented above provide a better indication of the cross section of the concrete cores.

Visual estimation of w/cm based upon the cement paste microstructure may also provide insights on the w/cm ratio. A difficulty with this approach is the inherently heterogeneous nature of the microstructure, so a number of regions were selected. Like the Danish procedure using fluorescent epoxy to estimate w/cm ratio, the SEM image porosity appears in contrast to the surrounding cement paste. In this case, we have three lab-produced mortars based upon the mix design of the deck concretes but prepared at w/cm of 0.30, 0.35, and 0.40 and cured for about 45 d. An example of this approach for the Echo Wash core and the three laboratory mortars is provided in Figure 35. Based on these images, it is difficult to assess the likely w/cm of the in-place concrete and thus, the point counting procedure presented above is the preferred method for estimating this value.

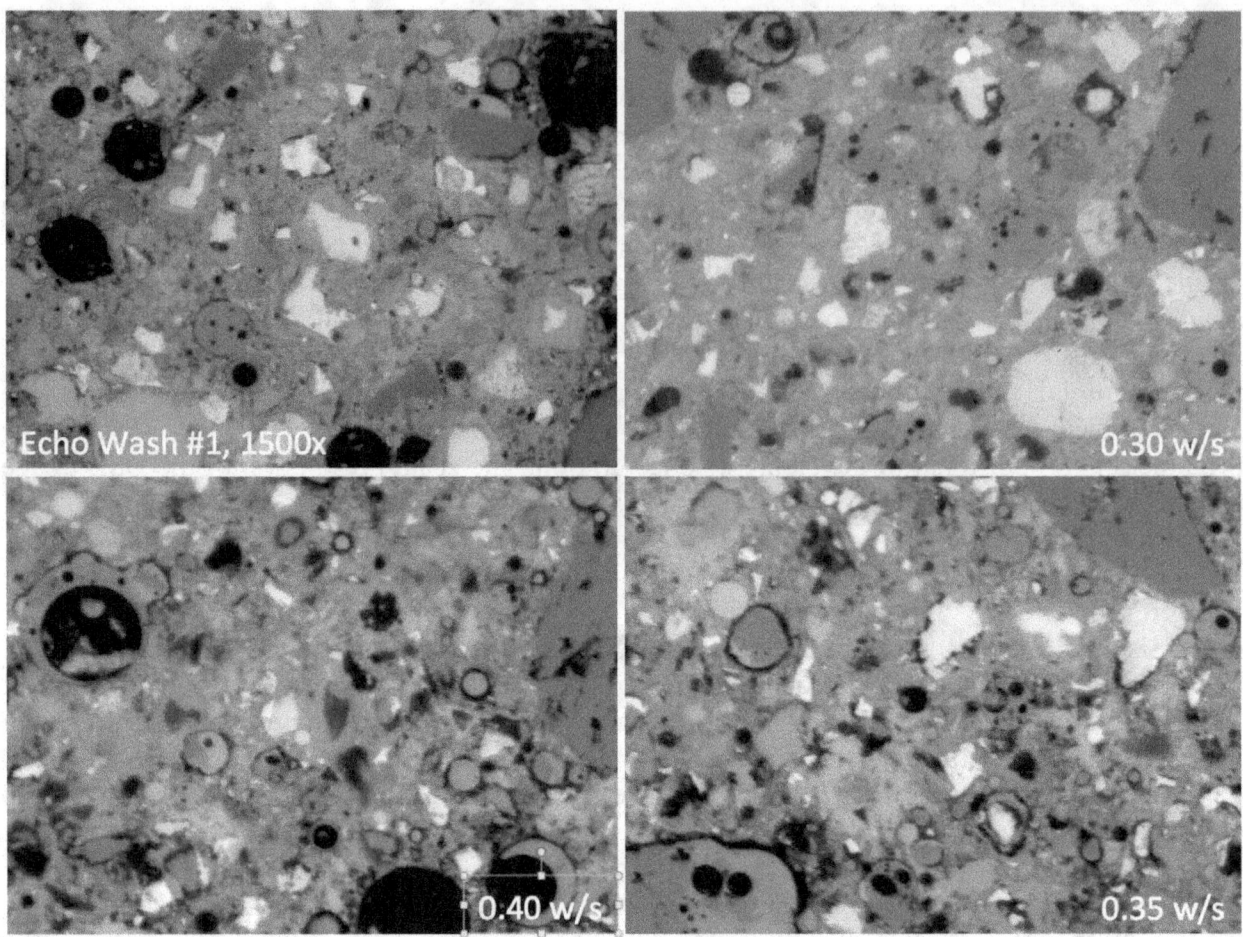

Figure 35. Echo Wash cement paste microstructure and comparison to mortars prepared at *w/cm* (or water-to-solids ratio, *w/s*) of 0.30, 0.35, and 0.40.

5. Synthesis of Tasks 1-3

A considerable amount of information has been collected in this study in support of Tasks 1, 2, and 3. The following observations are provided to focus the synthesis of these results:

1) Experimental measurements of autogenous and drying shrinkage have indicated that autogenous shrinkage increases significantly as the *w/cm* is lowered from 0.40 to 0.30, while drying shrinkage varies little among the four mortars prepared in this study.

2) Autogenous shrinkage curves for a pure portland cement system typically level off at ages beyond 28 d. However, for the mortars produced in this study, autogenous shrinkage continues to increase at a significant rate at ages to 56 d and beyond. It is hypothesized that this is due to the pozzolanic reactions between the fly ash and the cement that are still quite active at these later ages, assuming that water is present in sufficient quantities.

3) A significant increase in drying shrinkage (on the order of 10 %) was observed when the mortar prisms were placed in a 25 % RH environment, following their 56 d exposure at 50 % RH.

4) The concrete microstructures of the cores provided by FHWA are very heterogeneous and the air content, while quite variable, is generally much higher than the values reported from the onsite testing of the fresh concretes.

5) Petrographic analysis and point counting indicates that the *w/cm* of the in-place concrete is likely closer to the specified value of 0.40 than the batched values of 0.31 to 0.32 for the Echo Wash and Valley of Fire concretes. This would be consistent with water being added prior to placement. Similarly, analysis of the Snake River cores indicated an in place *w/cm* of approximately 0.39.

6) There is also significant microstructural evidence of this retempering at the job site, particularly for the Echo Wash and Valley of Fire concretes, that further increased the microstructural heterogeneity and air content of the in-place concrete, consistent with the information provided by the Federal Highway Administration on this topic [20].

Table 9 provides the estimated total shrinkage at 56 d for the various mortars obtained by adding their 3 d autogenous shrinkage to the subsequent drying shrinkage obtained out to 56 d. The values are fairly similar, even though the autogenous (3 d) component of this total shrinkage varies from 2.7 % for the *w/cm* = 0.40 mortar to 15 % for the *w/cm* = 0.30 mortars.

Table 9. Estimated total (56 d) shrinkage for the four mortar mixtures.

Mortar	Total 56 d shrinkage (microstrain)
w/cm = 0.40	-715
w/cm = 0.35	-773
w/cm = 0.30	-723
w/cm = 0.30 no pigment	-693

As with many cracking scenarios, it is likely that there were multiple contributing factors to the cracking that ultimately occurred in these bridge decks. While the laboratory mortar testing indicates that a reduction in *w/cm* would likely increase the propensity for cracking (or at least not reduce it), the petrographic analysis has suggested that the best estimate for the *w/cm* of the in-place concretes is near their specified value of 0.40. Based on the batch tickets, this would be consistent with water being added at the job site. Given that the *w/cm* was near 0.40, and given that one would not expect such severe cracking at this *w/cm* value, what factors most likely contributed to the cracking? The current hypothesis includes the following:

1) The high air content of the in-place concrete and the heterogeneity (potentially produced by retempering, etc.) of the microstructure, both with respect to variability in air content and in density of hydration products, suggests a microstructure where water could be drawn from more porous areas to denser areas during the hydration (curing) process. This would produce less hydration in these more porous areas that would compound any inherently weak regions produced by the variable and high air content.

2) The high reactivity of the fly ash at ages of 7 d and beyond (see Figure 5) produces significant increases in both chemical and autogenous shrinkage, which could contribute to an increased propensity for cracking even months later. While the window for early age cracking for conventional portland cement concretes may be during the first month of curing, that for mixtures with significant quantities of fly ash as replacement for cement could be several months later [21].

3) The weather conditions present on the bridge decks in the months of January to March of 2010 may have been the final contribution to the cracking for the Echo Wash and Valley of Fire bridge decks. A rainy period in January would have provided additional moisture for the hydration and pozzolanic reactions, further refining the pore structure to produce smaller pores that will generate larger stresses if subsequently emptied. Then, the extremely low RH in February and March (ultimately down to 20 %) would have indeed emptied these pores during drying, producing very high stresses within the concrete microstructure and ultimately the observed cracking.

There are a variety of mitigation strategies available to reduce early age cracking [3, 4]. Two strategies that have been developed in the laboratory and successfully applied in the field are shrinkage-reducing admixtures (SRAs) and internal curing using pre-wetted lightweight fine aggregates (LWFA). SRAs generally reduce the surface tension of the pore solution (by up to a factor of about 2) so that less autogenous stresses are generated within the pore solution under sealed curing conditions, resulting in less autogenous shrinkage. They also delay the drying process so that measured drying shrinkages are typically reduced by as much as 50 %. SRAs have been available commercially for over 10 years from numerous chemical admixture suppliers.

Internal curing provides a sacrificial source of internal water in relatively large pores. The emptying of these larger pores within the LWFA will generate much lower stresses than those produced when the much smaller pores within the hydrating cement paste empty. As the cement paste hydrates or dries, its smaller pores imbibe water preferentially from the larger pores within the LWFA to maintain saturated conditions within the cement paste. This significantly reduces the stresses and strains produced either during sealed curing or exposure to drying. Through their individual mechanisms, these two mitigation strategies each produce a significant reduction in early-age cracking tendencies [3]. Internal curing has been employed in bridge decks in New York and Indiana, as well as in pavements and a large railway transit yard in Texas [22]. An ongoing research project sponsored by the Oregon DoT is evaluating both internal curing and SRAs for achieving shrinkage reduction in bridge deck concretes.

6. Conclusions and Future Research

The evaluation of laboratory-prepared mortars and petrographic analysis of field concrete cores have indicated several likely contributors to the early-age cracking observed on three bridge decks in Nevada and Wyoming, including their abnormally high air contents, consistent with the retempering of the fresh concrete with additional water, and the drying stresses and strains that were subsequently produced in the hardened concrete as the ambient relative humidity fell below 30 % in Nevada in February to March of 2010. While the additional water did not exceed that permitted according to the mixture specifications in each case, the energy of mixing applied in the field was likely insufficient to properly re-mix the concrete, as the produced concrete exhibited a quite heterogeneous microstructure and an extremely high air content, including a substantial quantity of entrapped (as opposed to entrained) air voids. These high air contents were not detected by either the evaluation of fresh air contents or the measurement of the compressive strengths of field cylinders, as the sample preparation procedures employed in producing these two types of specimens apparently removed much of the entrapped air that remained in place for the actual cast-in-place and hardened concretes. For the bridge decks in Nevada, the early-age cracking occurred during a dry period in February and March of 2010, with average daily ambient relative humidities below 30 %, which followed a relatively rainy period in January. These low ambient relative humidities would produce additional drying stresses and strains that could induce the observed transverse cracking patterns.

This study has identified two major topics that could be subjects of future research. The first would be the development of a test method to identify retempering of fresh concrete in the field before it has been placed. The present study indicated that current tests for cylinder strength and fresh air content may not be adequate for this purpose. The second topic for future research would be an examination of the influence of air content on drying shrinkage and restrained cracking. While it is well known that increased air contents significantly decrease compressive strength, their effects on drying shrinkage and cracking are less well documented. A controlled study evaluating the drying shrinkage and restrained ring (ASTM C1581 [16]) testing of mortars and/or concretes of similar mixture proportions, but of differing air contents, would provide a valuable reference database addressing this open question.

7. Acknowledgements

The authors would like to thank Mr. Max Peltz of the Engineering Laboratory for his assistance with material characterization.

8. References

1. Concrete Foundations Association of North America. "Concrete Cracking," 2011 [cited 2011 6 October 2011]; Available from: www.cfwalls.org/foundations/cracking.htm.
2. American Concrete Institute, "Causes, Evaluation, and Repair of Cracks in Concrete Structures," ACI 224.1R-07, American Concrete Institute, Farmington Hills, MI, 22 pp., March 2007.
3. American Concrete Institute, "Report on Early-Age Cracking: Causes, Measurement, and Mitigation," ACI 231R-10, American Concrete Institute, Farmington Hills, MI, 46 pp., January 2010.
4. Bentz, D.P., and Jensen, O.M., "Mitigation Strategies for Autogenous Shrinkage Cracking," *Cement and Concrete Composites*, **26** (6), 677-685, 2004.
5. Holt, E.E., "Early Age Autogenous Shrinkage of Concrete," in *VTT Building and Transport*, Technical Research Centre of Finland, 2001.
6. Sakulich, A.R., "Reinforced Geopolymer Composites for Enhanced Material Greenness and Durability," *Sustainable Cities and Society*, 2011(In Press).
7. Emmons, P.H. and D.J. Sordyl, "The State of the Concrete Repair Industry, and a Vision for Its Future," *Concrete Repair Bulletin*, 7-15, July/August 2006.
8. American Society of Civil Engineers. "2009 Report Card for America's Infrastructure," [cited June 20, 2010]; Available from: www.infrastructurereportcard.org.
9. U.S. Department of Transportation. "Deficient Bridges by State and Highway System," 2010 [cited October 10, 2011]; Available from: www.fhwa.dot.goc/bridge/deficient.cfm.
10. Electronic documents provided by Federal Highway Administration Central Federal Lands Highway Division to NIST, June 2011.
11. Miller, P.K., Y. Tinkey, and L.D. Olson, "Non-destructive Testing and Evaluation Investigation Concrete Bridge Deck of Echo Wash Bridge Lake Mead National Recreation Area Central Federal Lands Highway Division Project Clark County near Overton, Nevada," 2010, Olson Engineering: Wheat Ridge, CO. p. 22.
12. Miller, M.M., Y. Tinkey, and L.D. Olson, "Non-destructive Testing and Evaluation Investigation Concrete Bridge Deck of Valley of Fire Bridge Lake Mead National Recreation Area Central Federal Lands Highway Division Project Clark County near Overton, Nevada," 2010, Olson Engineering: Wheat Ridge, CO.
13. Rothstein, D., "Petrographic Investigation of Concrete Cores Extracted from the Valley of Fire and Echo Wash Bridge Decks Located at Lake Mead National Recreation Area, Nevada," 2010, DRP Consulting: Boulder, CO.
14. U.S. Department of the Interior, Bureau of Reclamation, *Concrete Mix Rules of Thumb*, available at www.usbr.gov/pmts/materials_lab/concrete/mixrules.pdf, accessed October 2011.
15. http://www.cemp.dri.edu/cgi-bin/cemp_stations.pl?stn=OVER&prod=0, accessed October 2011.
16. ASTM Annual Book of Standards, Vol. 04.01 Cement; Lime; Gypsum. ASTM International, West Conshohocken, PA, 2010.
17. Bentz, D.P., and Turpin, R., "Potential Applications of Phase Change Materials in Concrete Technology," *Cement and Concrete Composites*, **29** (7), 527-532, 2007.
18. Bentz, D.P., "Internal Curing of High Performance Blended Cement Mortars," *ACI Materials Journal*, **104** (4), 408-414, 2007.

19.   Bentz, D.P., Peltz, M.A., and Winpigler, J., "Early-Age Properties of Cement-Based Materials: II. Influence of Water-to-Cement Ratio," *ASCE Journal of Materials in Civil Engineering*, **21** (9), 512-517, 2009.

20.   Federal Highway Administration, Petrographic Methods of Examining Hardened Concrete: A Petrographic Manual. Appendix C. Retempering, http://www.fhwa.dot.gov/pavement/pccp/pubs/04150/appendc.cfm, accessed November 1, 2011.

21.   de la Varga, I., Castro, J., Bentz, D.P., and Weiss, W.J., "Application of Internal Curing for Mixtures Containing High Volumes of Fly Ash," submitted to Cement and Concrete Composites, 2011.

22.   Bentz, D.P., and Weiss, W.J., "Internal Curing: A 2010 State-of-the-Art Review," NISTIR 7765, U.S. Department of Commerce, February, 2011.